phonebook

Henrietta Thompson

phonebook

A HANDY GUIDE TO THE WORLD'S FAVOURITE INVENTION

With over 500 colour illustrations

Thames & Hudson

On the cover, front, from left to right: Motorola DynaTAC model 8000S,
1984 (© Motorola, Inc.); Nokia 3310 (© Nokia, 2005); Motorola V70
(© Motorola, Inc.); KDDI Talby (© 2003–5 KDDI Corporation, all rights
reserved); back, top: Motorola DynaTAC promotional image, 1973
(© Motorola, Inc.); bottom: DynaTAC phone user, Hong Kong, 1985
(© Motorola, Inc.)

Designed by Charlie Sorrel

First published in the United Kingdom in 2005
by Thames & Hudson Ltd, 181A High Holborn, London WC1V 7QX

www.thamesandhudson.com

British Library Cataloguing-in-Publication Data
A catalogue record for this book is available
from the British Library

ISBN-13: 978-0-500-51254-8
ISBN-10: 0-500-51254-X

Printed and bound in China by SNP Leefung Printers Limited

Contents

Introduction 6

Part 1 **Culture Focus** 14

Where are you? 16
North America 22
Europe 30
Africa 44
The Middle East 50
Asia 56
Australasia 70

Who are you? 76
Fashion Focus 84
Generation Focus 92

Part 2 **Design Focus** 96

A History of Handset Design 98
Old Predictions 102
50 Most influential Handset Designs 108
Possible Futures 184
When Gadgets Become Extinct 194
The Environmental Impact 206

Part 3 **Technology Focus** 212

What's Inside? 214
Case Study: Samsung Hera 216
Case Study: KDDI au Project 222
Case Study: Vertu 228
Case Study: Nokia Fashion 232
Case Study: Motorola V3 RAZR 236
Case Study: Sony Ericsson Smart Phones 242
Text Messaging 246
Connectivity 252
Ringtones 260
Camera Phones 264
The Rise of 3G 270

Directory 278
SMS Dictionary 280
Technical Glossary 283
Acknowledgments 286
Index 287

Introduction

Phone technology moves so fast that publishing a book about phones can seem like a rather silly exercise. Before it even hits the shelves there's bound to be a major new twist to the tale – a new product launch, a new application, a new technological revolution – that renders it out of date.

But this attitude assumes that mobile phones still exist as a phenomena. You thought they did? Where have you been? None of the manufacturers – Nokia, Motorola, Sony Ericsson, KDDI, NTT DoCoMo – talk about making phones any more. Phone functionality is just an add-on feature for hand-held games consoles, navigation units, televisions or PCs. Now they all make mobile 'devices'.

From fixed lines to carphones to transportables, portables and eventually pocketables, the astonishing technological advances and fast-paced market forces mean that we have reached a turning point in phone-design history: the protocol for the form a phone should take has completely exploded.

It is, therefore, about time that the history of the mobile phone – as a stand-alone product that has changed billions of lives all over the world – was written. If the car was the constant by which the first half of the twentieth century could measure its design and social history, today's equivalent is, without doubt, the mobile telephone. No other product has witnessed a design evolution as radical, determinedly progressive and technologically unpredictable as the phone.

Regardless of how other new devices enrich our lives, throughout its evolution to eventual market saturation, the phone has had the greatest impact on the world's population. The ability to make and receive phone calls from anywhere, along with the entirely new means of communication that is text messaging, has revolutionized society.

The word revolution is overused and usually an exaggeration, but in this case it's an understatement and should be pluralized. Phones have changed

the way we keep in touch with friends and family, how we work and the speed at which we get things done. According to research by Intervoice in 2005, people are now so dependent on their phones that one in three people questioned were concerned that losing their phone would mean losing their friends. It is clear, too, that they have changed the way we talk: it is no longer 'Hi, how are you?', it is 'Hi, where are you?' Ringtones, meanwhile, have changed the soundtrack of our cities.

But that's just the beginning. Phones have changed the way we vote and express political dissent: mass protests can be converged in a matter of hours. They have changed the way we pray, from Muslims whose phones help them to locate the direction of Mecca to Catholics who receive daily prayers by text from the Vatican. Falling in love, too, is easier with phones, as the sweet language of SMS gets steamy, and people conduct relationships via a phone with people they'd never dare to meet.

left and right Advertising for the Nokia 7260 in 2004 showed that phones were not just useful but glamorous too

Everywhere you look phones have had an impact. In developing countries they offer a quick and easy way to leapfrog the slow and expensive process of building fixed-line networks. Often access to long-distance communication has made a life-saving difference; they have improved the way aid is distributed in times of need and healthcare has also been transformed by remote possibility.

Of course, not all the changes have been for the better. Phones are one of the most common contraband items in prisons all over the world, and the potential for their misuse by terrorists is substantial.

In part one, *Phone Book* looks at the impact this tiny – occasionally irritating – invention has had on culture around the world. Part two addresses the major design issues surrounding the phone. With the *50 Most Influential Handset Designs*, we see the evolution of its form from jumbo brick to spy camera and investigate how a *Star Tre*

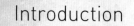

right Bluetooth-enabled beanie, Motorola and Burton, 20

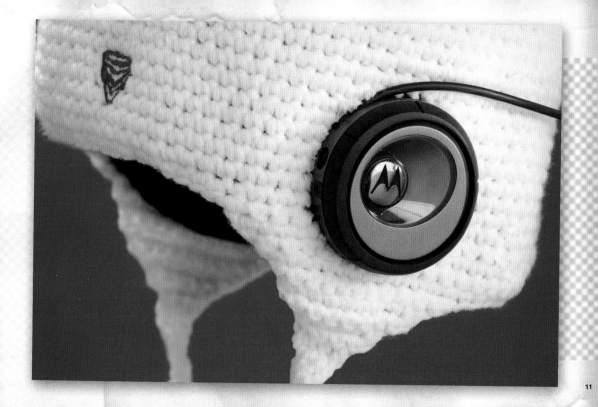

Introduction

prop informed the design of one of the most popular Motorola models of the early 1990s. And what of the future? Hold on to your Bluetooth-enabled beanies because there are plenty more changes to come, and if it is folly to document the fast-moving phone phenomenon at all, then let's do it properly and throw a little happy crystal-ball gazing into the mix in *Possible Futures*.

The focus in part three is technical: what actually goes into the device formerly known as a phone?

How are they designed and made, and how did we get it so wrong when we said text messaging would never catch on? We examine the technological developments that have made the phone indispensable: from text messaging to camera phones, connectivity to 3G, and all the ringtones in-between.

The phone is the world's favourite invention: let's celebr8 it ;-).

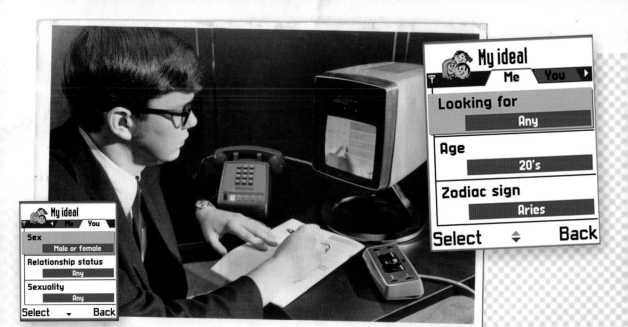

Part 1 **Culture Focus**

Where are you? 16

North America 22

Europe 30

Africa 44

The Middle East 50

Asia 56

Australasia 70

Who are you? 76

Fashion Focus 84

Generation Focus 92

Where are you?

Mobile, cell phone, hand phone…cuddly toy? At least according to the Swedes who nicknamed the device *nalle* – Swedish for teddy bear.

A standard terminology has evolved for the phone that sees variations on mobile, cell or hand phone used in most languages, but the Swedish *nalle* stands head-and-furry-shoulders above them all. The name originally referred to the term *yuppie nalle* coined because only rich yuppies could afford them and they often carried their phones as if they were mascots. Now *nalle* is only used to describe the phones of people who always carry them around and feel insecure if they misplace them.

The normal name for non-*nalle* phones in Sweden is more conventional – simply *mobiltelefon* or *mobil*. Like 'mobile phone' in much of the English-speaking world, including the UK and Australia, *telefon mobil* in Romanian, *telemóvel* in Portuguese, *mobiltelefon* in Hungarian and *mobitel* in Croatian, it's self-explanatory. Likewise, the Dutch call it

mobile cell phone keitai telefonino
nalle mobiltelefon mobitel telemovél
mobiele telefoon móvil mobilni
mobilos komórka selyular
Handy kännykkä portable
cep telefonu pelephone
farsími trubka

Where are you?

biele telefoon, the Spanish *móvil*, the Serbs *mobilni* and in Denmark they are *mobilos*. As in Sweden, Norway calls them *mobiltelefon*, and the Arabic term is pronounced *mobeel foon*.

Then there are names based on 'cell'. An American term that was coined to describe phones that use a cellular network, it – and variations on the theme – has been adopted globally. The Spanish sometimes say *cel* to make a change from *móvil* and the Polish *komórka* means the same thing. The Philippines's official word is spelled *selyular*, but usually they just say Nokia, which is fine, although confusing if the phone is, say, made by Motorola.

In many Asian countries they are called 'hand phones'. The Chinese term can be broken down into symbols for 'hand' and 'carry' (together meaning 'portable') and 'electricity' and 'speech' (together, 'a telephone'). The German word

Handy is unusual in that, while clearly related to the German word *Hand*, the 'y' ending gives the impression that it is an English word incorporated into German, even though the word does not mean phone in English. The Finnish *kännykkä* is another variation on the hand theme. Nokia trademarked the name in 1987 but it has since become a generic term.

In other places it's all about going the distance. The Japanese term *keitai* translates as 'to carry'. Sri Lankan phones are usually just called mobiles – but the real Sinhala word is *jangama durakathanaya – jangama* being 'something you can carry' and the other bit being the phone. The French *portable* likewise means just that.

'Pocket' phones are also popular, such as the Turkish term *cep telefonu*. In the native language of New Zealand, meanwhile,

19

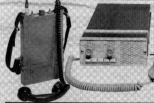

it's a *waea pukoro* – *waea* from the English word 'wire', and *pukoro* meaning 'pocket'.

But there are, like the Swedes, a few strands of independent thought. In Belgium, phones are generally known as *GSMs* (meaning Global System for Mobile communications) and in Iceland, while *farsími* is the official term, GSM phones are also known as *gemsi*, which, incidentally, also means 'young sheep'.

In Israel, *pelephone* literally means wonder-phone, and in Italy *telefonino* is small phone. In Russia, phones are often called *trubka* ('tube' in English) because the receiver part of stationary phones was called telephone tube, and now that's all they are.

But, for the purposes of this book, I will just be calling them 'phones'.

Although cellular technology was first developed and introduced in North America, cell phones initially made less of an impact in the United States than they did in Asia or Europe. This was partly because of a slow transition from analogue to digital technology and partly because cell phone services and connections were extremely expensive for a long time, with customers requiring impeccable credit ratings to get a phone. But the most prohibitive factor was the lack of need: fixed phones were everywhere already.

Eventually they started to take off, and now cell phones are common everywhere in the United States. But Americans still have a love/hate relationship with them. In 2004, a Massachusetts Institute of Technology survey placed the cell phone top of the list of most hated inventions that Americans couldn't live without. Phone manners have been much discussed. Unbridled use of phones in public can cause great offence, and inconsiderate use is not happily tolerated.

It's a stark contrast to Japan, where people love their phones as if they were their pets. But as phones have got more prolific, America – never a country to do things by halves – has gone one step further with phones for pets. Developed in Arizona, PetsCell was the first phone for dogs and cats, allowing owners to talk to their pets while away from home.

SHHH!

" THE REST OF US
DON'T CARE
WHAT HE SHE SAID TO YOU."

SHHH! | Society for HandHeld Hushing

"INSIDE"
VOICES,
PLEASE

SHHH! | Society for HandHeld Hushing

THE WORLD IS A
NOISY PLACE.
YOU AREN'T
HELPING THINGS.

SHHH! | Society for HandHeld Hushing

DEAR CELL PHONE USER

We are aware that your ongoing conversation

about,

your big night out

is very important to you, but we thought you'd like to know
that it *doesn't interest us in the least.* In fact, your babbling
disregard for others is more than a little annoying.

This message brought to you
by a concerned member of:

SHHH! | Society for HandHeld Hushing

e Society for Hand Held Hushing (SHHH!) issues these cards
emphasize the importance of phone etiquette

There are some interesting differences between how Americans and rest of the world use their phones. In most places young urbanites were the first to use the new technology, and as a result the push has been for younger, more streetwise functions and styles. In the United States, however, it was driven at the outset by a middle-aged, middle-class and business-user customer base. Texting was never the phenomena it was elsewhere, and at one point people were complaining that phones were getting too small.

Push-to-talk technology was also big news stateside, while it's not been a success globally, and people often use phones as if they were walkie-talkies – holding the phones to their ear to listen, and to their mouth to talk.

Motorola's *Beep Box* film in 2004 showed exciting new potentials
push-to-talk technology

Canadians, just like their closest neighbours in the United States, didn't take to phones as fast as those in Europe and Asia. However, the technology – especially that for mobile applications other than talking – is being heavily promoted to a demanding and increasingly phone-savvy youth market.

One idea to encourage Canadians to have more phone fun was Say Hello Toronto. In 2004, Nokia installed jumbo screens in downtown Toronto and its representatives persuaded passers-by to have their picture taken (with the Nokia 3220 or 6225 camera phones, which were being plugged at the time), and then posted the images onto the electronic billboards.

Although penetration remains lower than other countries, Canadians are not too far behind when it comes to technological innovation. One 2004 initiative called Murmur adapted the idea of 'hanging' text messages

that were popular in Japan at the time. A Murmur logo,
a phone number and a location code were posted in
various places around Toronto, Vancouver and Montreal.
Anyone passing one of these locations could call the
number and listen to a commentary about it. The headphone
sets available at particularly touristy museums suddenly
seemed prehistoric.

416.915.6877
213611

murmur
www.murmurtoronto.ca

The first ever mobile phone call in the UK was made by comedian Ernie Wise on 1 January 1985 across the Vodafone network.

Within twenty years, ninety per cent of the population owned at least one phone, and it had become the fastest-selling consumer product. The UK is one of the world's most mature mobile-communication markets. It also has the highest bills.

The phone can be held partly responsible for the British losing their stereotypical dour, umbrella-wielding, bad-at-flirting reputation. Walk down any city street and you'll hear all sorts of emotional turmoil and personal outpourings at very audible volume, though it's not always well received by eavesdroppers. Phones have replaced umbrellas as the most frequently turned-in item of lost property, and one-time national hero David Beckham ruined his clean-cut reputation with a text-sex affair.

In London, phones even provided a new way to hail a black cab, something that hadn't changed in more than

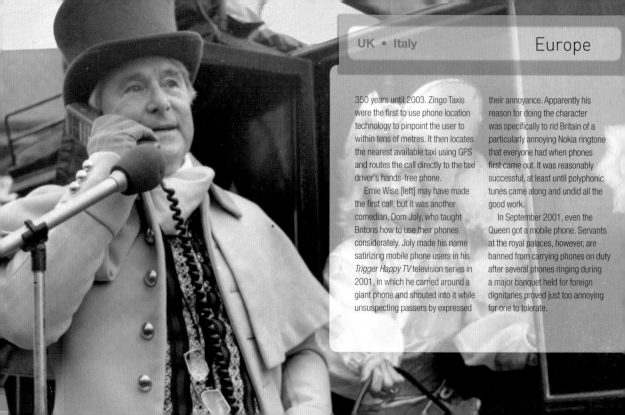

350 years until 2003. Zingo Taxis were the first to use phone location technology to pinpoint the user to within tens of metres. It then locates the nearest available taxi using GPS and routes the call directly to the taxi driver's hands-free phone.

Ernie Wise [left] may have made the first call, but it was another comedian, Dom Joly, who taught Britons how to use their phones considerately. Joly made his name satirizing mobile phone users in his *Trigger Happy TV* television series in 2001, in which he carried around a giant phone and shouted into it while unsuspecting passers by expressed their annoyance. Apparently his reason for doing the character was specifically to rid Britain of a particularly annoying Nokia ringtone that everyone had when phones first came out. It was reasonably successful, at least until polyphonic tunes came along and undid all the good work.

In September 2001, even the Queen got a mobile phone. Servants at the royal palaces, however, are banned from carrying phones on duty after several phones ringing during a major banquet held for foreign dignitaries proved just too annoying for one to tolerate.

Although Italians aren't always very keen on new technology, often preferring more traditional ways of getting things done, they have taken to mobile phones like Americans took to pizza.

Italy had adopted the *telefonino* with gusto by the early 1990s. Helped by the introduction of pre-paid phonecards (Europe's first) in 1996, phones quickly became affordable and accessible to most people.

The love affair with phones can, perhaps, be attributed to the culture's sociable nature. A high importance is placed on socializing, and Italians like to talk. In contrast with the comparatively inhibited peoples of the UK and America, in Italy there is no stigma about talking on the *telefonino* in public. If a person is making a video phone call in a restaurant, for example, an appreciative crowd will gather in a matter of seconds, watching and even joining in.

Phones have been incorporated into Italian culture to a great extent: multimedia messaging was a much bigger hit in Italy than in the rest of Europe, and Telecom Italia Mobile was the first network to supply soccer video highlights to its customers. Italians can also enjoy receiving a daily text message with prayers from the Vatican if they want.

Thrifty young Italians, meanwhile, have developed a new way of using phones to communicate. The *squillo* is a system based on not answering your phone when it rings. *Squillare* means to ring, and a *squillo* is a one-ring phone call.

The phone records who the call is from, but it is not meant to be answered. Instead it's the long distance equivalent of a smile.

Depending on the context it can mean different things. If you are running late to meet someone, you might send one to say 'I know, but I'm on my way' or if you're the waiting party you might send one to remind the person not to forget about you. You might send a *squillo* because you want someone to call you back, or it can be a way to let someone know that you like them. Get someone's phone number from a mutual friend and *squillo* them every day until they *squillo* you back. Ingenious.

Most commonly a *squillo* simply says '*ti penso*' ('I'm thinking about you'), and it's free.

In 2004, the Italian left was outraged after 56 million people received a text message from Prime Minister Silvio Berlusconi (many in the middle of the night) urging them to turn out to vote in the next day's European and local elections.

'Clearly, the ownership of three television channels and political control of the RAI [Italy's public broadcasting network] are no longer enough for [Mr Berlusconi],' one centre-left MP, Roberto Giachetti said. Mass texting was authorised in Italian law only 'in cases of disaster or natural calamities' and 'for reasons of public order or public health and hygiene'. Berlusconi's people argued that in this case the messages would ensure a steadier flow of voters and thus avert any threat to public order.

You are sitting in a meeting and your phone rings. The first instinct of many Europeans would be to cancel the call and turn the phone off. In Spain, however, and especially in cities, the priority is to be available. The Spanish don't use voicemail as often as in other countries – instead the onus is with the receiver of the call, even if they are in company.

Phones have changed life in Spain, and not just as an excuse to interrupt meetings. Text messaging has transformed its political landscape. Saturday, 13 March 2004, the night before the general election, became known as 'the night of the mobile telephone' when a text message spread among a newly disillusioned electorate made thousands of people congregate within a few hours in front of ruling-party Partido Popular's headquarters.

The message called for two things: to meet at 6 p.m. on calle Genova and to find out the truth.

The government was widely believed to be staging a cover-up regarding the perpetrators of the terrorist train bombings in Madrid two days earlier. The day before an election, political protests are forbidden in Spain, and any organization behind one could be punished. In this instance, however, there was no single organization responsible.

According to operators Vodafone and Amena, twenty per cent more text messages than usual were sent that day. Never before had a protest been created so fast, with so little warning and had such an influence. The government was ousted the next day.

The French fiercely guard their culture and way of life from the potential intrusion of *le portable*. France was the first country in Europe to pass a law that allowed restaurants, theatres, cinemas and other public places to install mobile-phone jammers, preventing people from making all calls except to emergency services. The ruling would ensure that the French could enjoy their escargot without interruptions.

Phones in France are otherwise popular, and most people, as in the rest of western Europe, own at least one. Outside of restaurants and theatres, phones are part of the culture.

They are even part of France's literature, and the country claims to be the first to publish a book written entirely in SMS slang. Aimed at teenagers and featuring an anti-smoking storyline, the novel is called *Pa SAge a TaBa* – French text speak for 'not wise to smoke'. At one point, the Dtektive (detective) asks the smoking villain: '*6 j t'aspRge d'O 2 kologne histoar 2 partaG le odeurs ke tu me fe subir?*' Expanded and translated by textually.org the passage reads: 'What if I spray you with cologne so you can share the smells you make me suffer?'

It's not exactly Proust, but teenagers with phones are smoking less anyway [see p. 94] so the whole idea may be redundant.

Apart from culture, another thing that the French are remarkably good at is protesting. Mobile phones were used to cultivate this skill further when, in 2004, many customers thought that French operators were charging too much for text messaging. Dissent ensued, petitions were signed, SMS strikes were organized and demands were made for refunds. Operators were forced to take notice.

Phil Marso

Frayeurs "SMS"

OS'KOUR! ALED!

DIKO "SMS" inclus

mégacom-ik EDITIONS

Phil Marso

Pa SAge a TaBa

v.o SMS

polar live

Not only did Germany invent the sport of phone throwing, but they are also exceedingly good at it. The idea emerged when German phone companies started offering cheap phones that broke too easily and, since the first championships were held in 2000, the sport has spread to many other countries.

As phones got smaller and lighter, the sport got more competitive, but Germans still seem to retain the edge. According to Nico Morawa, the 2004 world record breaker (throwing his Siemens 67.5 metres), the skill is in choosing the right technology as well as having the physical strength to lob it. The best method, he says, is to throw the phone as if it were a discus.

Germans are skilled in more than one way when it comes to phones: they are also the acknowledged world champions at texting.

The national penchant for SMS is sometimes thought to be a response to the patchy telecommunications system in the country that existed prior to reunification of East and West in 1990. In the former East Germany many people had no individual phone service at all, while in the West, the state monopoly system had off-puttingly high rates. Many Germans, therefore, skipped the fixed-line stage altogether and bought a *Handy* at the first realistic opportunity. Text messaging was billed at a flat rate and was the cheapest – and most popular – option.

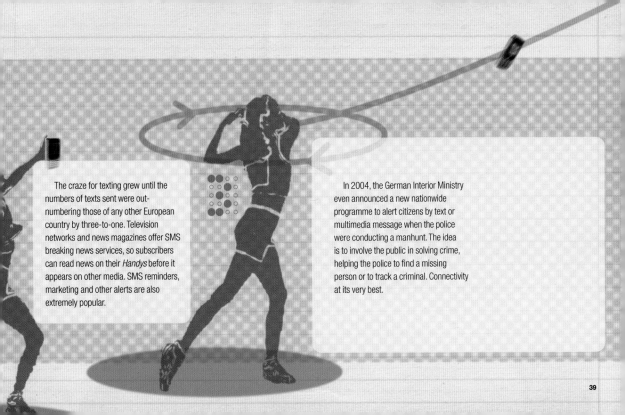

The craze for texting grew until the numbers of texts sent were out-numbering those of any other European country by three-to-one. Television networks and news magazines offer SMS breaking news services, so subscribers can read news on their *Handys* before it appears on other media. SMS reminders, marketing and other alerts are also extremely popular.

In 2004, the German Interior Ministry even announced a new nationwide programme to alert citizens by text or multimedia message when the police were conducting a manhunt. The idea is to involve the public in solving crime, helping the police to find a missing person or to track a criminal. Connectivity at its very best.

Finland and phones go together like reindeer on rye. Finland has topped the mobile-phone league since phones were first invented, and one of the reasons for this is Nokia.

Nokia was named after the town it was founded in, which lies 200 kilometres from Helsinki. Although it started out by making rubber tyres and paper, it is now the world's largest producer of phones, controlling a large chunk of the global market and selling more than 100-million handsets every year.

Like the rest of Scandinavia, Finland has a long history of excellence in design, combining practicality, functionality and aesthetics in a way that few countries manage. Nokia is no exception, and much of the company's global success can be attributed to the design of its phones.

Finnish people are justifiably proud of the company, and will correct bemused foreigners if they admit that they always assumed that Nokia was Japanese.

Nokia might symbolize the phone revolution in Finland, but there are other reasons why the Nordic country took to them so well too.

Keeping in touch has always been paramount amongst Finland's widely scattered population. Having information is key to surviving the extreme weather. Some attribute Finnish love of phones to the appeal of communicating at arm's length rather than face-to-face. Despite a reputation for being dour and introverted, and despite their deep respect for silence, Finnish people still need to talk.

The popularity of phones has been encouraged further by reasonable prices. In the early 1980s, the Finnish government deregulated the telecommunications sector, creating a competitive market place and stimulating a fierce price war.

One of the most wired and wireless nations in Europe, the phone industry and the internet were largely responsible for the Swedish economic boom of the 1990s, and saw the capital city Stockholm become newly hip, with a young, techno-savvy population.

Swedish people love their phones, and it's not just because the partly Swedish company Sony Ericsson makes such good ones. Communication has long been crucial in Sweden – in 1900 Stockholm had more phones than London or Berlin, and the Swedes took to the internet and wireless technology with similar enthusiasm. As in Finland, a small population scattered over an inhospitable landscape has a great deal to do with it, but Sweden is bigger than its next-door

neighbour and has more industrial and financial muscle.

By 2004, phone subscriptions had surpassed the number of inhabitants as many people have one phone for personal use and another for work.

Sweden was also one of the first places in the world to implement a strategy of m-government. As most of the population has at least one phone, it made perfect sense for public bodies to exploit the technology. The national tax authority, customs agency and several local authorities all communicate with customers by text message, while a scheme called Gateway Sweden issues lorry drivers with customs clearance by SMS, saving hours.

Stockholm city government has set up a range of mobile services under the banner

'mCity', which enables schools to manage truancy, care workers to manage rosters, and students to stay up to date with lectures and any last-minute bargains they might want to take advantage of. It also allows drivers to receive traffic information on their phones.

Sony Ericsson, P800, 2003

Add text
Add sound
Add recording

Russia had to wait much longer than many other countries for the phone revolution, but by 2004 it was the fastest-growing market in the world. Like the internet, phones began to appear in the mainstream at the end of the 1990s; before then they had been the preserve of wealthy businessmen. From 1999 to 2001 the number of phone owners grew five-fold as network fees fell dramatically. Today most Russians own a mobile phone, and, like many countries, the most enthusiastic phone users in Russia are the young.

While further west telephony had been a basic service that was taken for granted, in Russia the effects of the Communist economy took time to fade. Much of Russia lacked a fixed-line telephone service because the Soviet regime had not provided the basic infrastructure efficiently.

Fed-up of waiting for things to improve, many Russians – particularly those from rural areas – saw mobile phones as a more reliable alternative. And as the country's economy strengthened and personal incomes grew, phones increasingly became a part of people's lives in the cities too.

Russia's Communists are among the world's political activists who thought to harness the power of text messaging (*esemeski*) in their campaigns.

At a party meeting outside Moscow in 2005, the Communist Party's deputy chairman Ivan Melnikov lauded the benefits of more guerilla propaganda methods such as texting and graffiti. Members 'could use telephones to send political jokes or rhymes, or attract attention to events – anything that motivates a person to send the message along to someone else,' he was shown saying on NTV television.

Where mobile phone technology is available in Africa, it is the most egalitarian means of remote communication there is. Just a few years after their introduction in most parts of Africa, the number of mobile phones far exceeded fixed-line connections, and more Africans have begun using phones since the beginning of the twenty-first century than in the whole of the previous century.

The phone has transformed many businesses and become a tool for economic empowerment. Farmers in Senegal, for example, share a phone and use it to connect directly with buyers in Dakar who will pay three-times the rate offered by local middlemen. Others follow hourly fluctuations in coffee and cocoa prices on their (shared) phone, selling their crops when world prices are most advantageous.

For most Africans, however, the cost of handsets and services are still out of reach. But, while only the rich can afford their own phone, poor communities simply share them. Although a phone may nominally belong to a single person, in Senegal, as with many other African countries, phone is often regarded as the property of the community

Sometimes, several families share a handset, and in places without electricity they will charge the phone from car batteries. Children run between neighbours when calls are expected or need to be made.

Whereas elsewhere in the world the phone is considere a highly individualistic and personal tool, in this context it is treated more like a fixed-line phone. By keeping it in one place it becomes useful for whole communities.

Another widely used money-saving trick is the 'beep-call'. Like the Italian *squillo* [p. 33], people can contact eac other by dialling the number and then ending the call befo it incurs charges. When it was discovered that one African phone company was only charging calls from the third second after a connection is established, users would hold whole conversations by calling each other in turn, speakin for less than two seconds and hanging up.

MTN villagePhone

MTN	400/=
UTL	450/=
Mango	450/=
Celtel	600/=
East Africa	1200/=
International	from 3000/=

Like many developing countries, Uganda's fixed-line network is abysmal, and the number of landlines was quickly overtaken by mobile phones. A poverty-stricken nation, in the past a Ugandan villager would often have had to travel a long way to make a call, which would be extremely expensive. The cellular phone service providers do a booming business and often help customers with services such as dialling numbers when they have problems (such as poor eyesight) and ensuring privacy when they make a call.

Many villages in Uganda can receive mobile phone signals, but the service and handsets are too expensive for residents to afford. With one innovative, brave and wonderful initiative, copied from one that had already proved successful in Bangladesh where the same problems existed, MTN Uganda improved access to communications services, reduced poverty and helped the local economy.

MTN villagePhone was a joint venture between MTN Uganda and the Grameen Foundation USA. Individuals living in impoverished rural areas could become villagePhone operators by taking a micro-loan (as little as US $230), to be repaid over the next year, to buy the phone equipment that could serve their whole community.

The equipment package included a wireless handset, a user manual and a SIM card that could be loaded with pre-paid airtime. They could also get a car battery or solar-power panel if there was no electricity where they lived and an antenna for areas where the MTN network could only be accessed with a boost.

South Africa can boast a modern, sophisticated, nation-wide telecommunications system that is among the best in Africa. Under Apartheid rule this infrastructure was

SIMpill medicine dispenser, which uses SMS to regulate medication, 2004

MTN Network Coverage

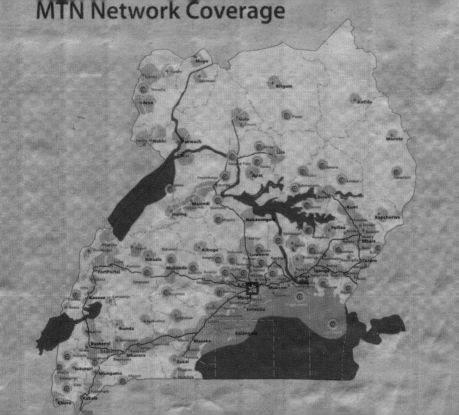

47

constructed primarily to serve the country's white population and was not extended into rural areas where the majority of South Africa's black population lived. By the time cellular networks were being established this situation had begun to be redressed, but the beauty of mobile phones is that, as long as there is coverage, they get everywhere and can be used by everyone.

South Africa now has better coverage for mobile phones than it does for terrestrial television, and the market for phones is a sophisticated one. One trend that seems to be unique to the country is the combination of hair salons and phone shops: literally, a hairdresser on one side

of a store and a mobile operation along the opposite wall. Apparently this is simply abou convenience: as phone shops don't need a l of space it is cheaper to share the rent with hairdressers or clothing shops.

Cape Town–based company SIMpill and Tellumat Communications have developed y another way to improve African's health usin SMS. Their innovation, the SIMpill bottle, del a text message to a central server whenever is opened. This can then be checked against patient's prescription instructions. If no mess is received SIMpill can then notify the patient a healthcare professional with a reminder.

To coincide with World Aids Day in 2004, World launched a new SMS service in Kenya promote awareness and knowledge of HIV a Aids. Subscribers to One World's free service could text questions about the disease and

Africa

ive a prompt answer. They could also opt
ceive daily tips on preventing infection and
ng others.

obile phones are one of the best means
mmunicating in Kenya since many more
le have access to a phone than to the
net, and text messages are popular
use they are easy to use and cheap.

The Middle East

People across the Middle East are enthusiastic and demanding consumers of the latest, smallest, coolest phones. But if you see someone taking out a phone in the street in Dubai, it may not be to make a call or check a message, but because his phone has just reminded him that it is time to pray.

LG Electronics's F7100 Qiblah phone – a benchmark in regional innovation – was the first of several phones developed specifically for followers of Islam. While it was already possible to sign-up for a text message service that would send a reminder of the five-daily times for Muslims to pray towards Mecca, LG went one further and, as well as an alarm, the handset featured a compass that would indicate the direction of Mecca.

Dubai-based Ilkone also produces phones that incorporate the call-to-prayer function, but have added the text of the Qur'an stored on its handset too.

Muslims are now well served when it comes to phones rich with Islamic features. Even those with regular models can (and do) augment them with a religious ringtone or download a lunar calendar.

In the conservative Arab state of Saudi Arabia, the emergence of camera phones caused some serious trouble. In September 2002, after reports that phones were being used to secretly photograph women, the chief of the Commission for the Promotion of Virtue and Prevention of Vice took decisive action and banned the phones.

It was a rather extreme reaction to the problem and it wasn't popular. Companies continued to advertise the phones on billboards and television. The black market thrived, and traders bought phones from neighbouring Gulf countries and sold them at a considerable mark-up.

Photography is generally banned in Saudi Arabia, and women are sometimes sensitive to being photographed without a veil. However, following the ban it was women who were seen to be breaking the law. One girl was expelled from her university for taking pictures of her

friends and posting them on the internet, and a woman was attacked by other (female) guests at a wedding for having a camera phone and using it.

Eventually the government realized that education might be better than prohibition at preventing the abuse of technology and the ban was lifted in December 2004.

Messaging, multimedia or not, often makes the headlines in Saudi Arabia. Phone contact between boys and girls is not acceptable but religious texting is very popular.

During Ramadan, messages reminding Saudis to practise their religious rituals or to undertake charity work are widely sent. When photo-messaging became legal again, some of the most commonly sent picture messages were images of the names of Allah and verses from the Qur'an.

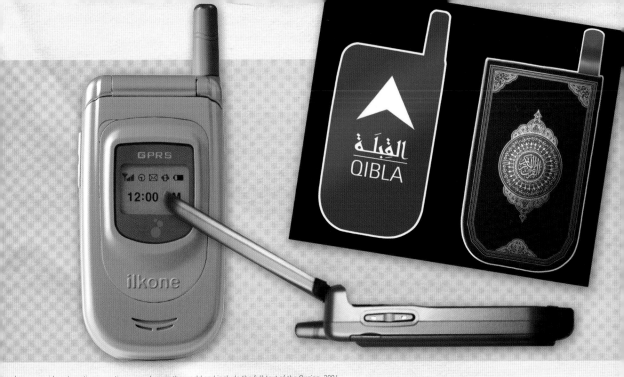

he phones provide automatic prayer times anywhere in the world and include the full text of the Qur'an, 2004

Chatty Israelis are well known for being fond of communication technology, and it should come as no surprise that the rate of adoption of phones in Israel was speedy. The audacious motor-mouth Jewish mother may be only a stereotype, but it doesn't mean she doesn't exist, and phones are the perfect way to spread gossip. Many Israelis use their phones constantly and unapologetically, indicating that there is a certain obligation to report on what is going on and how they are, wherever they are – even in places where phone use is prohibited, such as restaurants, theatres and trains.

Israelis have no qualms about speaking at length, even long distance, and tend not to view the cost as a barrier to communication, especially if someone else is footing the bill (phones are often given to employees as a tax-deductible expense).

Along with Pelephone and Cellcom, Orange is one of the major service providers in Israel, with a thriving business- and lifestyle-user base

On a more serious note, Israel's political climate has also had an impact on the use of phones. With terrorism rife, and the fear of it always present, the need to communicate is paramount; people need to be able to check-up on their loved ones at short notice. While the networks haven't always been able to cope with the sharp increases in traffic after a major event, phones offer the best peace of mind it is possible to get.

The Israeli term for mobile phone, *pelephone*, which is also the name of one of Israel's largest operators, combines 'telephone' with the Hebrew word *pele*, meaning wonder.

Not only are phones – *keitai* – ubiquitous in Japan, they are taking over. In Japan's cities you will see a vast number of people talking, shopping, reading emails, sending text messages, dating, browsing the web and playing games on their phones. Whether they are queuing, drinking in a bar or travelling on the subway, from toddlers to old age pensioners, they are all at it.

Walking around Tokyo without a phone would be like walking barefoot. It's much the same situation as in South Korea, and between them the two countries have the best cellular service and the biggest choice of phones in the world.

Although influenced by American and European trends and tastes, Japanese phones are more technologically advanced than their western counterparts.

While the rest of the world plays catch-up, phones in Japan have developed some unique forms too, such as the Wristomo phone – a watch that folds out into a handset from NTT DoCoMo and Seiko.

In 1999, when the rest of the world was wondering what wonders WAP would bring, Japanese phones were already capable of accessing the internet and sending and receiving email with NTT DoCoMo's i-Mode, a long time ahead of similar services in Europe. Other companies followed and created original Internet content for mobile-phone users. It worked well and has been described as a 'second internet'. Shopping by phone took on a new meaning: this was the m-commerce revolution.

The Japanese are well known for their love of gadgetry. They are also known for their

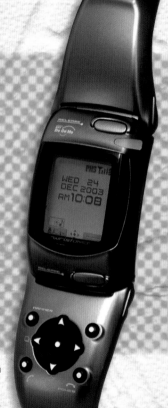

NTT DoCoMo and Seiko's Wristomo, 2003

consumerism and penchant for cute things. There are a lot of very cute phones in Japan, but more than that, there are a lot of cute accessories. Phones in Japan are important identity statements: hanging stuff off them and sticking stuff on them – straps, antenna rings, fake gems, flashing LEDs – is widespread. Even adults do it. Even important businessmen and politicians do it. The *keitai* is a way to establish a personal identity within a well-defined group.

One step on from Nokia's version of customization, Xpress-on covers, Panasonic Mobile Communications created flat, screw-on faceplates for its phones, which could be fashioned and customized by anyone, from big-name designers to independent craftsmen. It went down a treat in a country where people get bored of their phones very quickly.

For visitors to Japan there is a free phone service that provides translations, road maps and alerts about cultural events around certain cities. One of the first countries to adopt RFID technology, which allows information to be attached to or incorporated in a product to be read by a receiver in another device – a phone, for example, RFID tags are now installed all over the place: walk around and point your phone at something to instantly receive more information about it.

NTT DoCoMo's Pocket Post Pet, featuring a dancing pink bear and other email delivering 'pets', was a craze in Japan in 2001

NTT DoCoMo's Eggy allowed users to play music and video
clips over wireless networks, 2002

In November 2002, 'hand phones' were introduced to North Korea, and in just one year 20,000 North Koreans had bought one. On 24 May 2004, phones were banned.

The North Korean government has a strict monopoly on information. It is one of the most culturally isolated countries in the world. To travel abroad without government permission is to commit treason, while reading prohibited publications, or listening to foreign radio broadcasts are serious crimes.

It is impossible for an economy to survive in the world without some access to modern technology, and oppressive regimes such as North Korea's struggle with the necessity to allow, even encourage, its use but to restrict it heavily at the same time. But it proved impossible to control the networks and mobile phones became a weapon for anti-government movements. Mobile phones were being used to make contact with neighbouring countries or, worse, America. The official reason given for banning them was that they could be used to trigger bombs in terrorist attacks.

It is rumoured that North Korea still has a mobile phone network in Pyongyang but it is only for use by government officials and a few VIPs.

Two more different phone cultures could not exist side-by-side as North and South Korea. The number of mobile-phone users in South Korea is predicted to reach 39.43 million, 81 per cent of the population, by 2008. They are prolific

IM-7400, a self-sterilizing phone for the Korean market, 2004

users of technology and people of all ages have phones and use them for surfing the internet, mobile gaming and for sending text and picture messages.

South Koreans are such keen internet users that some of their web portals boast some of the highest traffic ratings in the world, even though their content is all in Korean. It also has one of the world's fastest mobile phone networks.

Like in North Korea, in the South phones are considered a powerful political weapon, but, in stark contrast, they are used to encourage democracy. In the presidential elections in 2002, the winning candidate, Roh Moo-hyun, had little need for mass rallies or traditional campaign tactics. Instead he sent text messages to almost 800,000 people urging them to vote. Much of the campaigning for and discussion about the elections were held on-line. In this land of 'e-politics', phones and the communication technologies they support are transforming the system.

Pantech GI100, a fingerprint recognition phone, 2004 [p. 180]

Samsung S250, an impressive five-megapixel
camera phone, 2004 [p. 160]

Mobile telephony was introduced to India in 1994, though it took a long time to take off. With only a few service providers, heavy regulations, huge licence fees, high call charges and expensive handsets, only the privileged could use a mobile phone in India.

Things have changed. Now, phone tariffs and handsets are some of the cheapest in the world and, although the numbers of people who use a phone are still lower than in Europe, they are no longer considered such a luxury.

India has a large population of under twenty-fives, and the cheap rates lure a huge youth market. In many urban areas phones are must-have gadgets, while others choose them simply because it helps business, reducing the need for expensive office space or travelling to clients.

In more rural parts of the country where the telecommunications infrastructure is poor, a mobile phone can be a lifeline. Unfortunately still too expensive to be accessible to many, one company dreamt up a solution. Shyam Telecom, which operates in Rajasthan, has established a service that makes phones more mobile than ever. It has equipped a fleet of rickshaws with phones and employed drivers to pedal them around the state capital, Jaipur, and the surrounding countryside.

The hand-pedalled tricycles are equipped with a battery, a billing machine and a printer. The calls are cheap and the drivers, usually disabled people and women (who are often dependent on their families for income and were often below the poverty line), receive commission.

Shyam Telecom also equips camels with wirelessly connected computers for use in the desert.

As well as having the world's largest population, China also has the biggest market for phones. This may seem surprising considering that landline and internet coverage for the country was so poor. But mobile communications, as elsewhere in the world, just served to resolve this difficulty. It grew at an alarming rate to become indispensable across society. No business, however small, can now survive without a phone, and China's tens-of-millions of migrant workers use them to search for jobs and keep in touch with their families in the countryside.

While the developed coastal areas are mature markets where people are demanding sophisticated gadgets, inland income levels are much lower. Still representing a huge chunk of China's customer base for phones, in these areas phones need to be cheaper: their basic function – being a phone – is most important.

Mobile networks are reaching even the most remote areas, where previously phones were only available in local government offices, often miles from many villages. The mobile phone has, in fact, bypassed the fixed phone in these areas to the extent that people have been installing desktop cell phones. It looks like an ordinary landline, but works on a GSM account, and is much cheaper.

In the cities, however, where phones are a social necessity, the market is driven more by style. The phone capability goes without saying (no one actually makes calls on phones any more do they?), so is less important than games, text-messaging interface, camera resolution and the device's appearance. It was also one of the first places to get a handset with an in-built television (the NEC N940; see p. 165).

and right NEC N910, a square handset, 2004

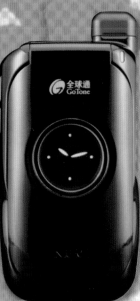
NEC N620, a handwriting recognition phone, 2004

Besides SIM-card vending machines and limited-edition phones that have been blessed by Chinese deities, one important innovation in the Chinese phone market has been the '*Cai Ling*' or 'colour ringback tone'. This, thanks to China Mobile, is the facility to personalize the answering tone – whether the phone is ringing, engaged or otherwise. When you dial a number in China you might hear personalized monologues, jokes and such-like replacing the beeps. It's a wonder why no one thought of it earlier – a feature being annoying hasn't stopped phone manufacturers before…

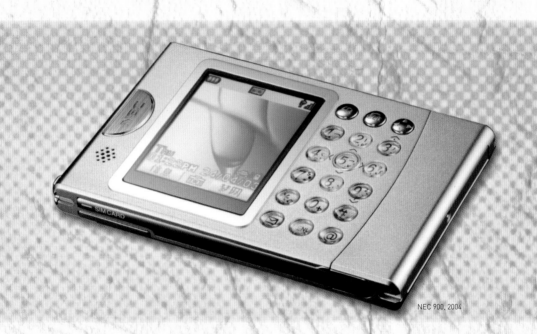

NEC 900, 2004

Australia is home to some of the oldest life forms, stromatolites, and the oldest people, yet has always been one of quickest to take-up new technology, especially mobile phones.

Australia has one of the highest penetration rates of phones in the world – the phone industry is now worth more than the car industry – and it is home to several 'firsts' brought about by phone use.

A Sydney man was arrested and fined, and his phone destroyed, after he was caught taking surreptitious photographs of topless women on Coogee Beach with his Nokia in 2004.

Just as progressive in the fight against phone corruption, Australia's Virgin Mobile launched a radical new service in the country in the same year to help its customers avoid making drunken phone calls.

A survey of 409 people by the company found that ninety-five per cent had made drunken phone calls. Of those calls, thirty per cent were to ex-partners, nineteen per cent to current partners and thirty-six per cent to other people, including their bosses.

By dialling a special number, plus the phone number they don't want to call when drunk, Virgin would kindly bar calls to it until 6 a.m. the following day.

But Aussies aren't all so badly behaved – far from it. Australia was also the first country to introduce the 'cellular samaritans' concept. These civic-minded citizens call radio stations to report traffic hazards and congestion, imminent storms, long queues and so on. Mobile phones, it seems, could make a serious contribution to their country's welfare, providing a means for people to engage in small acts of social responsibility.

Phones are, in fact, seemingly used in every aspect of Australian life, from paying for parking (Australia was one of the first countries to trial the m-parking concept) to checking-in for a flight. Phones are even helping fruit and vegetable growers find people to pick their crops. In 2003 the Federal Department of Employment teamed up with the National Harvest Labour Information Service in a scheme to text unemployed people about where workers were wanted.

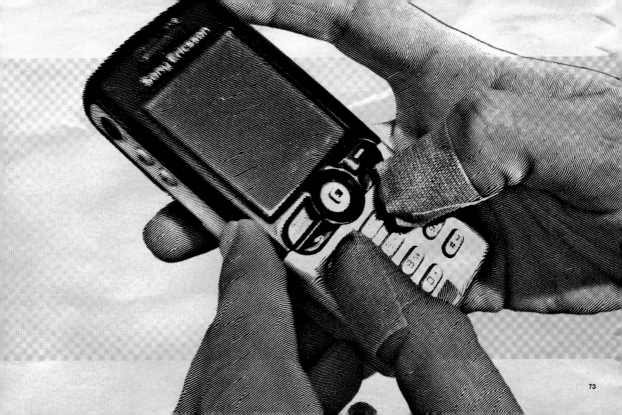

Australasia

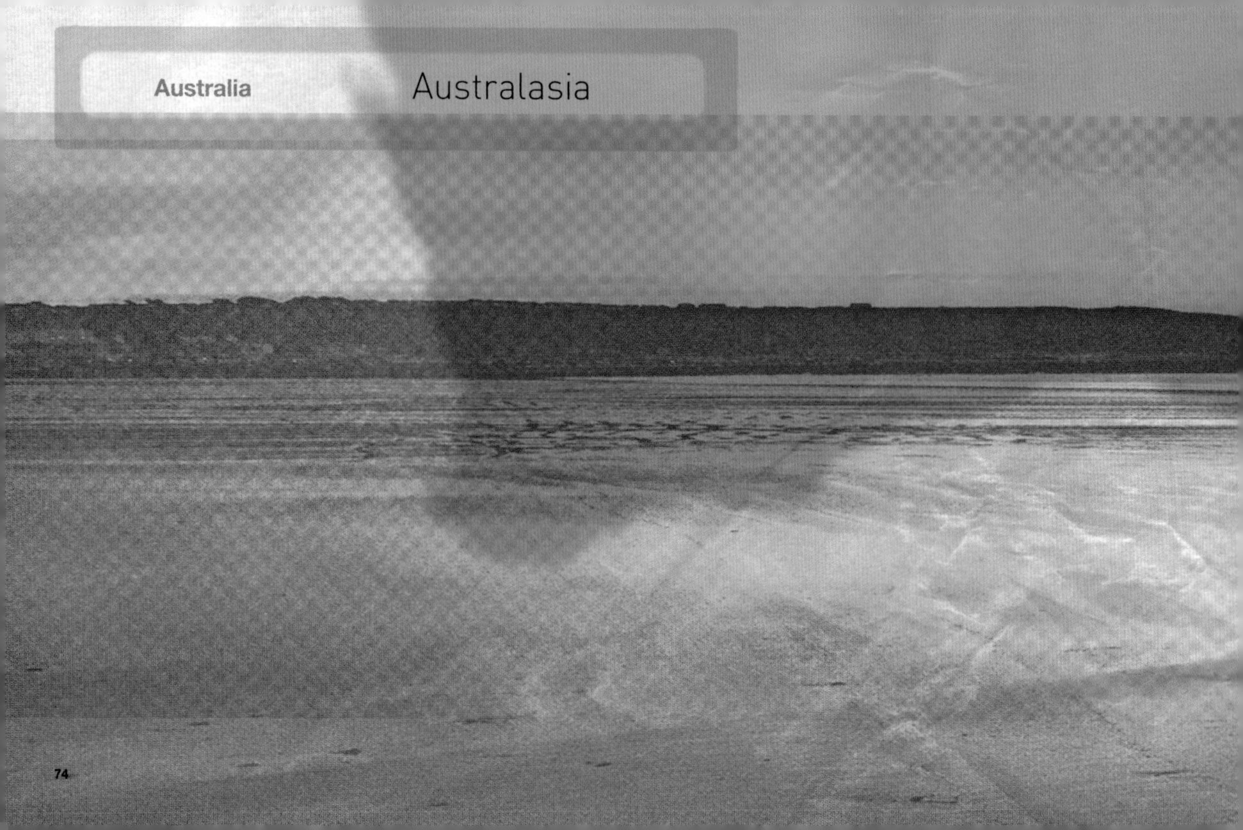

Another milestone in Australia's phone history was the phenomenon of injury by too much texting. Only a year after the National Day of Safe Text (23 July 2003), a thirteen-year-old girl was diagnosed with texting tendinitis. The doctor's report was published in the Medical Journal of Australia, alongside a call to phone manufacturers to place prominent warnings on packaging about possible injuries caused by overuse. The girl's SMS habit had caused a painful, two-centimetre swelling of the tendon on her forearm. Also a keen knitter, it had taken her doctor a while to pinpoint the source of the problem, but knitting uses both hands, and she was only using one thumb to type out the messages.

Who Are You?

With many countries now exceeding a hundred per cent market saturation, you would think that the question of who is most likely to own a phone would be easy to answer. Everyone.

It is not quite the case yet. Although global markets are developing fast, and mobile networks are bringing telephony to more people in developing countries than was ever possible with fixed-line phones alone, for every person who owns more than one phone there are groups of people sharing one between them.

Meanwhile, there are still people who don't have any need for phones – just not very many of them – and not being able to get hold of them, of course, we don't know who they are. There are also pockets of the world, such as North Korea, where phones are banned.

Looking at the phones on offer today, and the new niche markets that are constantly being discovered or invented in an attempt to sell more phones, it is hard to see any gaps that are left to fill. There are phones for babies that act as alternative babycom systems, phones for pets – paw operated collar-mounted gadgets that enable little fido to hear his master's voice even when they are apart. There are

Who Are You?

big-buttoned phones for the elderly, crystal-encrusted ones with concierge services included for the very rich, and for 'ironic' fashion leaders there are customized retro eighties phones too. Phones have taken the concept of niche to new heights: there is, should you be paranoid, a phone that alerts its user if he or she has bad breath. And, if you really have nothing better to do with yourself, there is a handset that measures air pollution.

Whatever turns you on, there's a phone for you. And that means, like it or not, that the model you own speaks volumes about you.

This wasn't always the case. Nokia, which launched forty new models in 2004, was once producing just four per year. Choice was limited and, as a direct consequence, so were customers. The phone companies had a tricky job selling products and services in a way that would appeal to everyone.

As with most new technologies, the early adopters were the young, the rich and the fashionable, and these markets continue to be the ones manufacturers want to attract when developing new designs and products. However, for a technology whose primary

Burton and Motorola,

Who Are You?

purpose is to bring people together, it is perhaps surprising how divisive phones can be.

The gap between the young, time-rich and curious and the old, technophobic and change-resistant continues to widen. There are also several significant sub-groups, including blind and deaf users, whose needs were for a long time either completely ignored or, in the case of the Nokia Communicator [p. 128], only addressed by accident. The etiquette of using phones, meanwhile, has caused conflicts on public transport, in theatres, restaurants and offices globally, while people learned – by trial and error – to cope with the new intrusion.

retro hands-free kit, 2004

Who Are You?

David Beckham in Vodafone live!'s promotional campaign, 2004

Ordinary people too could use the new technology, according to the campaign's television advertisements, 2004

Fashion Focus

As consumer electronics go, phones are perhaps the most personal. We carry them around with us and use them to send and store intimate messages. We keep the names, numbers and photos of all our friends and family on them. We show them off at parties and we let them vibrate in our pockets.

The phone is tied more closely to the individual than, say, a laptop or an MP3 player, and the make and model of a phone says a lot about who we are and what we do.

There are still many more types of people in the world than there are types of phone (just). To make a mass-produced, mass-market product convey the subtle nuances of our complex personalities, customization was the only answer.

It began in Europe with Nokia's Xpress-on covers and in Japan with accessories such as straps, antennae rings and stickers. Fashion houses such as Hermès, Gucci and Louis Vuitton all saw the potential for a quick buck and added mobile-phone straps and mascots to their collections.

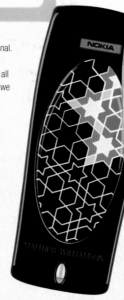

Xpress-on covers for Nokia designed by Matthew Willamson

Fashion Focus

The marriage of fashion and phones saw dollar signs light up for many phone manufacturers, but appealing to a fashion market was often harder to achieve than they thought.

One of the biggest mistakes, perfectly exemplified by Siemens with its disastrous Xelibri range [p. 144], was to assume that fashion was simply about frivolous colours and funky shapes. Siemens, along with Nokia and other manufacturers, made phones to be worn as if they were a weird, futuristic new type of jewelry.

But people who like fashion aren't always stupid, and just because a phone looked different, even if it looked great, it didn't make it cool.

Manufacturers learned the hard way that for a phone to be really fashionable, it had to function as well as anything else on the market, so a more popular tack was to re-fashion existing phones that were already selling well. Celebrities and fashion designers were brought in to endorse or

Phone straps by Berlin Diva,

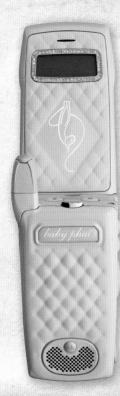

left Customized Samsung phone and arm
by Diane von Furstenberg,
above Siemens Escada SL55 model,
opposite David Beckham phones home for Vodafone live!,

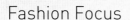

customize phones that would then be sold as special limited editions at a special marked-up price.

Nokia, one of the first manufacturers to really take fashion seriously as a design strand, still saw the future of phones as fashion accessories. Going beyond customization, this meant developing phones that looked good, related to trends on the catwalks and the high street, worked well and, crucially, were really easily interchangeable. Nokia envisaged people using different phones in different situations. One with PDA functionality and email for work; a tiny camera phone for posh parties; and one for holidays, perhaps with a television incorporated and a better camera. As products converge and as they get cheaper, this, say some manufacturers, is the future.

Fashion Focus

Now that devices are being embedded into our clothes, the real fusion of phones and fashion has arrived. In 2005, a collaboration between Motorola and Burton Snowboards saw Bluetooth-enabled jackets, helmets and beanies in a pioneering move to keep snowboarders connected, even when in mid-air and up a mountain.

Fashion designers and phone manufacturers must continue to work together to ensure communication technologies remain as stylish as possible in the years ahead. Because there's no doubt about it, from Bluetooth helmets to personal display systems on sunglasses, phones are getting more personal all the time.

Bluetooth-enabled snowboarding gear by Motorola and Burton, 2005

Generation Focus

Never mind 2G and 3G, the biggest generation gap in phones is between the people who use them.

In most countries it is sixteen- to twenty-four-year olds that drive the market. They are the early adopters and the most demanding of novelty. They are, therefore, the biggest spenders when it comes to new technology and the people manufacturers are most preoccupied with.

Mainstream users follow. Teenagers have the energy and time to try things that older users don't have the patience for, and so teenagers define what's hot and what's not. Manufacturers watch them and adapt new products for everyone else accordingly.

A person who has never known life without the internet and who does their school work on a wireless personal gadget is going to find perpetual contact easier to grasp than someone who has quite happily used libraries for the last fifty years. The young and the old have very different needs when it comes to phones, but it was only once phones were ubiquitous, and competition increased, th[...] phone companies turned their attentions to niche mark[...] such as different age groups. They found that older markets, particularly those over sixty, buy less phones a[...] use them less, but have much more loyalty, sometimes even keeping a handset for well over a year. They don't text much, but neither will they start to panic if they real[...]

right LG Electronics's KP8400, or diabetes phone,
which can test blood sugar levels, 2005

y've left the house without their phone.
Manufacturers such as LG and Kyocera [p. 167]
covered a demand among older users for simplicity
hones that don't look as if they might crash
parably or beep embarrassingly if they press the
ng button. Phones on which they could see what
buttons did and what it said on the screen without
ing to change glasses.

Around the same time another new market was
erging: the under tens. Young children were crying
for phones that were just as simple as the ones
older generations, but in a more exciting range of
ours. It would seem that even toddlers had some
for real-time connectivity, always-on information
on-the-move opportunities. From GPS tracking
ces to baby monitors, parents knew there was
ty in cells.

oung people, however, remain the biggest
rs for technology, or maybe it's the other
around. The ubiquity of phones amongst

Generation Focus

teenagers, meanwhile, has had some major effects on their lives. They have changed the way they study and the way they act in school (surreptitious texting overtook note-passing in class years ago), they have extended their social circles considerably and phones are used for everything from flirting to bullying.

It was even discovered that phones provided an unexpected health benefit – in 2000 the British Medical Journal reported that phones were driving down the rates of teenage smoking. The proportion of fifteen-year-olds who smoked fell from thirty per cent in 1996 to twenty-three per cent in 1999, and from thirteen per cent to nine per cent amongst eleven- to fifteen-year olds. Phone vouchers, it seemed, were offering effective competition to cigarettes in the pocket-money stakes, offering adult style, sociability and rebellion all in one, healthier package.

As if giving-up smoking wasn't forward-thinking and innovative enough, it was the teenagers and twenty-somethings who invented texting. They asked for more games, more functions and better looks. Teenagers will decide when email is dead (in South Korea they have already) and when wearables are really wearable.

We may have to wait a little longer, therefore, for phones to be built into zimmer frames.

MyMo phone for babies, 2004

Nokia Hello Kitty phone, 2004

A History of Handset Design   98
Old Predictions   102
50 Most Influential Handset Designs   108
Possible Futures   184
When Gadgets Become Extinct   194
The Environmental Impact   206

2000

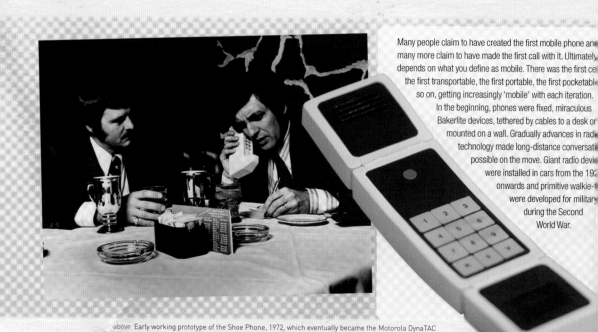

Many people claim to have created the first mobile phone and many more claim to have made the first call with it. Ultimately, depends on what you define as mobile. There was the first ce the first transportable, the first portable, the first pocketable so on, getting increasingly 'mobile' with each iteration. In the beginning, phones were fixed, miraculous Bakerlite devices, tethered by cables to a desk or mounted on a wall. Gradually advances in radi technology made long-distance conversati possible on the move. Giant radio devic were installed in cars from the 192 onwards and primitive walkie-t were developed for militar during the Second World War.

above Early working prototype of the Shoe Phone, 1972, which eventually became the Motorola DynaTAC
right Concept model for a 'double-flip' phone, 1972, an early version of the Motorola DynaTAC

artin Cooper, a project manager at Motorola, is widely
ted with inventing the first personal, hand-held phone,
ynaTAC, in 1973. It was smaller than anyone had
ght was possible. It was huge.

ter a decade of publicity and network-building efforts,
er's DynaTAC was launched in 1984. Before that, the
ransportables – luggables – lent an odd air of glamour
se rich enough to afford them.

unsuspecting public slowly realized the benefits
his new technology could bring, although initially
were outweighed by difficulties. The phones were
st bulky, they were extremely expensive to buy and
e. As battery power was minimal, they came with a
y-pack suitcase. Tales of plugging the phone into a
cigarette lighter and watching the headlights dim are
ondly, as are stories of calling one from a phone box
vatching money disappear faster than it was possible
k coins through the slot. They were great, however,
nergencies, for changing plans at the last minute, for
conversations that mostly consisted of 'I'll call you
on a landline' and for showing off.

These advantages were compelling, and the idea slowly
took off. As the cellular networks grew, phones began to
shrink. Thanks to advances in silicon-chip and battery
technology handsets quickly slimmed down. The number
of suppliers grew and their marketing began to appeal to
consumers' sense of style as well as their wallets.

For a long time it was all about miniaturization. Portable
became pocketable and pocketable became wallet-sized.
Manufacturers embarked on a race that saw handsets
become positively anorexic: the slimmer the phone the
more stylish it was perceived to be.

The different approaches to shrinkage resulted in a
variety of shapes, many of which went on to provide design
standards. Motorola's innovative flip design was a strong
favourite, while the clamshell was useful in protecting the
screen when closed and extending ergonomically to form
the mouthpiece when open. Nokia favoured the candybar
shape, which was simple and logical, and allowed for easier
customization too.

As phones became smaller, cheaper and ubiquitous, they
were increasingly designed to appeal to different markets.

Manufacturers experimented with colours and with new
materials, and the form of phones evolved away from that
of a fixed-line handset, as designers realized mobile phones
could be any shape they wanted to make them.

Nokia launched Xpress-on covers in 1998, marking a
new era in phone design. Electronics had never been so
personal, and entire industries grew up that no one would
have imagined could exist a decade before. Flashing
antennae, downloadable screen characters and novelty
ringtones all served an increasingly avid fanbase.

The evolution of phone design was subsequently
shaped by two forces: the growing importance of personal
expression through consumer electronics (including phones
as fashion items); and advances in handset technology.

At the start of the twenty-first century, the variety of
handsets expanded dramatically as phones began to
incorporate colour screens and built-in cameras. The size
and resolution of screens became paramount, and being
able to input data was as important as receiving it, creating
a need for innovation in keypad design and touch screens
too. Meanwhile, designers were let loose to experiment with

form, resulting in swivel phones, 'jack-knife' styles, rotating screens and slide mechanisms. Crucially, having spent two decades making devices smaller, suddenly they needed to get bigger. The rules of the game had changed.

As more new functions emerged, new markets came with them. Phones divided into more categories, each further removed from the original telephone than ever before. Traditional voice-centric handsets were overtaken by multifunctional gadgets and niche products that were task specific, such as Nokia's N-Gage [p. 164] and the BlackBerry [p. 136]. These were often aimed at users for whom telephony was a secondary function. As bandwidth increased, phones became capable of far greater things. With the advent of 3G, manufacturers finally abandoned the word 'phone' altogether. The all-encompassing term 'device' was more appropriate.

Still not necessarily meaning one product that does everything, a new notion emerged: the personal mobile gateway. With a modular approach, the idea was for users to carry two or three products – such as a mobile phone, a music player and a television – each with their different specifications of screen size and sound quality and all linking to a network using a Bluetooth connection to a central unit. As people started to own more devices for different functions, times of day or handbag size, products were designed to be interchangeable. The Nokia 7280 [p. 174] was a good example of this, with an easily ejected SIM card making up for any limitations posed by its small camera-concentrated form.

One of the biggest constraints designers have to contend with is the battery. As fuel cell technology edges closer to reality, the design of the devices looks set to change dramatically. Without a battery compartment there is more freedom to play around with the positioning of other components, such as the display and keypad.

One part of the phone has evolved almost independently from the handset; if the development of phones has taught

Promotional image of Motorola's 'mobile' phone, 1973

us anything about technology, it is the importance of a good user interface. Nokia dominated this area for years, contributing to its position as number one in the phone company charts for so long. But for years little research was conducted into what makes a good interface – one reason why an application like texting was originally overlooked – and we are yet to see any serious innovation in the field.

So handsets it is. The next chapter is devoted to the evolution of phone design and includes some of the most influential handsets in the development of mobile devices. See what people thought would be the future of phones before the mobile phone came along. Laugh at how wrong they were, and then find out what kinds of predictions are being made today. Finally, reminisce about the prehistoric landline phone and see what other products we'll soon say goodbye to, as the all-conquering mobile phone marches on.

The 'transportable' Mobira/Nokia Talkman from the early 1980s preceded the slightly smaller Cityman [p. 118]

'One of these days you won't even recog[nise]
your telephone,' predicted Robert A. Kelly[,]
writing in *Science and Mechanics* magazi[ne]
1961. In the future you could 'make and re[ceive]
a call anywhere – from your car, or even wa[lking]
down the street', or 'transmit business infor[mation]
automatically and at high speed over telepho[ne]
lines'. Just imagine!

It was all very exciting. But we've not always[been]
so good at predicting how design and technolo[gy]
will evolve. For example, for years the idea of see[ing]
the person you were talking to (in colour!) was th[e]
next logical step for phone companies to take. Ph[ones]
would incorporate a picture screen specifically for [that]
purpose, while 'electric eyes' would transmit your i[mage]
to the person at the other end of the line. Eventually,[just]
before the idea became properly feasible, people dec[ided]
it wasn't such a good idea. Dashing out of the shower[...]
to answer the phone would be a far more complicated[...]
business, while taking a sickie from work or having an [...]

Old Predictions

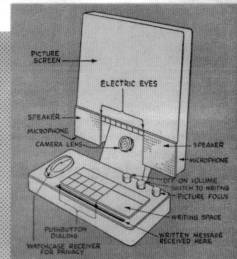

PICTURE
SCREEN

ELECTRIC EYES

SPEAKER
MICROPHONE
CAMERA LENS

SPEAKER
MICROPHONE

OFF ON VOLUME
SWITCH TO WRITING
PICTURE FOCUS

WRITING SPACE

PUSHBUTTON
DIALING

WRITTEN MESSAGE
RECEIVED HERE

WATCHCASE RECEIVER
FOR PRIVACY

Old Predictions

would require backdrops, make-up and a whole series of props. We collectively decided that perpetual contact was a great achievement for modern science, but only if we retained complete control over it. We were wrong about so many things....

Text messaging was a big surprise, for example. SMS technology was originally intended for behind-the-scenes functions, for network engineers to communicate with each other. Nothing more. It was extremely confusing for the phone industry, therefore, when customers hijacked the idea and used it for personal communications, developed a new language for it and demanded new services to support the innovation. The phone companies had previously assumed that extended SMS communication would prove too difficult for widespread adoption.

When data services first took off, it was predicted that phones would offer wireless internet: the world in our pockets. Of course they did, and do, but because of device limitations (essentially they are too small) this has not happened as expected. Web pages are not designed to fit on a phone display no matter

how big it is. Most operators realized that they were unable to offer a full internet service, and devised their own portals to give consumers access to content, rather than the generic information they get online.

And then there was WAP. Wireless Application Protocol – brilliantly named to make it as inexplicable and inaccessible as possible – never quite happened because it was so hideously misrepresented by companies trying to sell it. We were told we could surf the web, and when we actually got a slow, unreliable approximation we were understandably disappointed. As a result, the WAP standard never became the same for phones as html was to the internet, like everyone had said it would. And the emergence of CDMA phones, which provide access to content portals rather than the full internet, made sure of it.

We changed our minds about how we wear our phones too. The logical conclusion to phones getting smaller and lighter is that eventually they would cease to be visible at all, which suggests that they would be part of our clothing. While wearable computing is becoming a reality, it's not how we expected it to be: keypads installed directly into jackets and wires threaded through for earphones and a mike. Instead we have a whole new meaning for the phrase 'smart clothes'.

While everyone knew that phones would have a big impact on how we keep in touch with each other, few believed that they would mean we would see more of each other. Ten years ago it was widely predicted that in the future the majority of human interaction would be done in the e-world, but, although we are getting a lot done virtually, we are not doing it alone. People are getting together more often and in larger groups, and it's all thanks to the wonders of telecommunication. Instead of replacing face-to-face relationships, it has enhanced them by making them happen more often and with more people.

Perhaps more than anything, though, we get the numbers wrong all the time. In 1990, forecasters were happily projecting

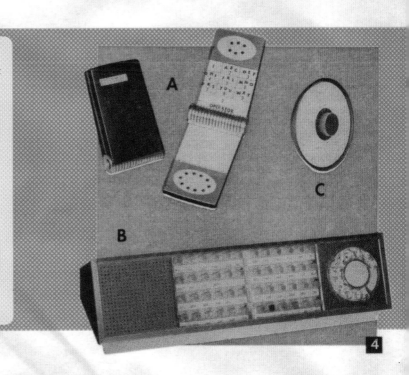

Old Predictions

that by the end of 2000 there would be 100-milli mobile-phone users in the world; the actual figu was 650 million subscribers to phone networks with 700,000 joining every day.

But it's not all unfounded conjecture. Even in the smallest space of time, and with vast resea programmes, we are still getting it wrong.

By 2000, companies were investing huge sums in research that they hoped would help them foresee the future for digital technologies. Samsung – at the time better known for silicon chips than phones – was one such manufacture

Samsung Design Europe's early ideas of what a 3G phone might look like were fun but inaccurate, 2000

Having dispatched its design teams to find out what people actually wanted from technology, Samsung eventually devised a radical new series of concept designs that everyone loved. Samsung combined futuristic 3G themes of convergence with the understanding that many people shirked technology, not making use of the features on offer because the futuristic shapes and poor interfaces left them alienated and technophobic. The designers' products solved this by using natural materials and the familiar forms of diaries, pens and make-up compacts.

Needless to say, the concepts remained just that as attitudes to technology change rapidly and total submersion made silver and black boxes and shiny rows of buttons more familiar than electronic make-up cases. And now we know: only time will ever tell.

Samsung's *Matrix* phone, 2003, went into limited production for hardcore fans of the film, but evolved into a more aesthetic form in the SCH-S250 [p. 166]

50 Most Influential Handset Designs

Motorola SCR-100 Star Trek Communicator
Motorola Transportable Dancall Motorola DynaTAC
Nokia Cityman Nokia 2110 Motorola StarTAC
Nokia 6110 Nokia 8110 Nokia 8810 Nokia 8850 Nokia 9000
Communicator Sony CMD Z1 Nokia 3210 Nokia 3310
Nokia 8210 Nokia 7110 IDEO Kiss Communicator
RIM BlackBerry Handspring Treo Vertu Signature
IDEO Social Mobiles Siemens Xelibri Motorola V70
Samsung T100 Sony Ericsson T300 KDDI Infobar
Siemens SL55/65 Orange SPV Nextel Push-to-Talk Motorola 3G
NEC 3G T-Mobile Sidekick Sharp Vodafone live! Sony Ericsson P800
Sony Ericsson T610 Nokia N-Gage NEC N940 Samsung S250
Kyocera TU-KA S Phone KDDI talby LG 8110 LG KP8400
Motorola V3 RAZR Nokia 7280 Isamu Sanada HiPod
Vodafone Toshiba V602T Pantech GI100 Samsung SCH-S310

Picking the milestone designs in the history of phones is difficult only in that everybody has an opinion about what should be on the list. And they can get very emotional about it too. This is because there is now a generation who grew up as the technology did and formed very strong attachments to their own choice of devices through the decades. If asked, product designers have a tendency to believe that the phones they designed were the landmarks from which all others followed. Likewise, people get extremely nostalgic about their first phones. But emotional allegiances have no place in a definitive and sober tome such as this one.

The fifty phones chosen for this section have been included because they fit certain criteria. They must in some way be a 'first'. Generally this is in terms of the handset's form and design, but some have had an impact for other reasons such as a technological advances or specific functions, for example the first 3G phones or the first motion-sensing phone. Secondly, regardless of whether they were ever produced or, in the case of the Star Trek Communicator, were designed as a phone at all, they had to have some form of legacy or lasting influence. This – as with the Siemens Xelibri range – didn't always have to be positive either, sometimes we learn as much from the mistakes made as from successes.

By 1936, Chicago-based Galvin Manufacturing Corporation had established a thriving business in car radios, and had been improving connectivity in America for almost a decade. Its founder, Paul Galvin, decided to take his family on a luxurious six-week tour of Europe, visiting Italy, Germany, Austria, France and the UK. It was to be his inspiration for developing one of the most important pieces of military equipment of the Second World War.

Galvin returned from Germany convinced that war was imminent. After some encouraging conversations with army heads, and with a deep recession snapping at the company's heels, he ordered his engineers to investigate how radio might be applied to the needs of the military, which at the time was using heavy and cumbersome back-pack radios.

Chief engineer Donald Mitchell believed he could improve on this and subsequently invented the Handie-Talkie. A two-way portable radio, its colossal proportions now make the giant bricks of mobile phones from the 1980s look like Lego in comparison. But the size of the Handie-Talkie, no bigger than a biscuit tin and weighing just over 2.25 kgs (5 lbs), was a huge achievement at the time. With a microphone, head antenna and self-contained batteries, it had a solid range of one mile and a potential range of three.

Mitchell achieved it by using aluminium (although magnesium would have been lighter, it was relatively new and unfamiliar, which could have caused all kinds of problems). Scores of working models were tested, rejected and improved on: another innovation was a black nickel coating that prevented the retractable antenna from giving the game away to enemy snipers. And by 1941 it was complete. Four years later 100,000 Handie-Talkie AM two-way radios were in use by the United States and allied armies. They were indispensable, and used by soldiers to report positions and call for support and supplies.

Ten years after Galvin's fateful European trip, he changed the company's name to Motorola. Originally coined for its car radio, the new name suggested sound in motion. It stuck.

when *Motorola Radio* comes home from war...

Star Trek Communicator

DOCKING COMMUNICATOR

Either the Star Trek Communicator was one of the most visionary props of science fiction, or the designers of subsequent clamshell phones were all under a subliminal Trekkie influence. At the time, the hand-held communication device, with its black plastic body and flip-up activator, speaker grill, indicator lights and function keys, was designed to look like a piece of twenty-third-century hardware. Just a few decades later, not only has the vision of hand-held communication devices become a reality, but today's phones are even smaller and, unless you're keen contact someone on the Enterprise, more useful too.

The original Communicator was the brainchild of well-respected prop designer Wah Chang – the man who created the special effects for the film *The Time Machine* and the original *Planet of the Apes* series. The design evolved from the toy walkie-talkies of the 1950s, but Chang's innovation was to add a flip-up wire mesh, which many of today's phones echo.

As rumour-mongering Trekkies would have it, the working, spinning moiré on the prop was run by an old analogue stopwatch: Chang removed the lens and cover from the watch and then mounted the bottom disk to the second hand, which on an analogue piece would tick every tenth of a second. The jerky retro effect can be seen on episodes with an occasional close up. The knobs were allegedly fashioned from the wheel hubs of a toy car.

DOCKING
COMMUNICATOR

By the 2340s a new generation communicator had evolved. Built into the metallic uniform insignia the 'combadge' was worn on the left breast, and activated by touch or voice. Will this prove as far-sighted as the original design? US firm Vocera created a wireless voice communicator in 2003 that is clipped to a jacket lapel. As in the television series, to contact someone you press the talk button on the lapel badge, say a name and you will be put through to that person. Working via a Wi-Fi network, the gadget is smaller than a phone and has proved popular in hospitals to make it easier for staff to communicate with each other.

Motorola Transportable

By the mid-1980s cell phones were slowly making the transition from military high-tech to real life. If you ask enough people about their first phone you will find someone who was avant-garde enough to own one of these: a bag phone. Not 'mobile' in the way we know phones today, these were 'transportables', which came in their own, huge, faux-leather carrying case. Because, let's face it, this handset was not going to fit in your pocket.

The bag was also the transceiver, the antenna and an optional battery. The handset connected to the bag by a long curly cord, although you could also use the cigarette lighter in your car to charge it, but, if you wanted to talk for longer than a minute, you needed to have a spare car battery. Sound quality was – well, it wasn't – and reception was rare too, but it was still impressive. Tales are still recounted of people with them being stopped by strangers who didn't understand why they were standing in a car park shouting into a big black box. 'I'm on the phone,' would be met with absolute incredulity. 'On the telephone??' 'Yes', long silence… 'Oh MAN!'

There was no doubt about it, this was very, very cool.

Dancall

Founded in 1980, the Danish manufacturer Dancall was one of the first companies to introduce a portable GSM phone to Europe. This one, instead of being a typical bag phone like Motorola's Transportable [opposite], came with a hard plastic box about the size of a small VCR with a sturdy handle and a shoulder strap.

The handset attached to the front with a button to release it. A slot on the end held a huge rectangular battery and a giant antenna protruded from the other side, although even with this the reception remained intermittent.

As well as making and receiving calls, the Dancall model also made a good weapon against potential muggers.

Motorola DynaTAC

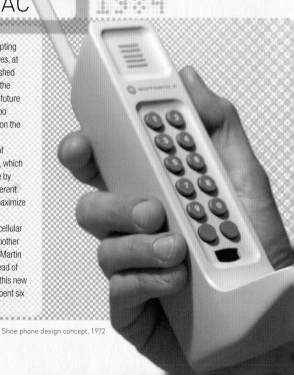

Witnessing how people had begun adopting portable radios and pagers into their lives, at the start of the 1970s Motorola established a mission for its product development: the ultimate communication device for the future couldn't be too small and couldn't be too light. And so the company began work on the DynaTAC.

By the end of the 1960s a network of small radio coverage areas called cells, which allowed multiple users to communicate by re-using the same radio channel in different cells, had begun to be established to maximize frequency use.

Anticipating the time when the first cellular networks would be launched (not for another decade, as it happened), their inventor Martin Cooper commissioned Rudy Krollop, head of design at Motorola, to determine what this new type of phone might look like. Krollop spent six

Shoe phone design concept, 1972

eks with a design team developing a series
product prototypes, and it was from these
t the basic DynaTAC – a phone that Motorola
oduced for the next decade in various
rations – evolved.

DynaTAC stood for the less catchy 'dynamic
aptive total area coverage', and everyone in
ollop's design team had a go at designing a
ncept for the handset. The double flip, the
nana, the retractable and the flip mouthpiece
ototypes were some of the most extraordinary
signs, and all have influenced phone design
ce, but only the 'shoe phone', as it was
ginally dubbed, by Ken Horn, was produced.

The DynaTAC emerged on the market ready
serve the new networks in 1984, a decade
ter Horn's shoe concept. There were few
anges to the original design except for an
tra column of buttons.

It certainly wasn't too small or too light,
but was impressive at the time. A 900 g
(2 lb) brick – its form appeared to be a direct
descendent of the Handie-Talkie, but instead of
warring soldiers, it was distributed among the
aggressively competitive business executives of
the 1980s.

Working prototype, 1973

DynaTAC 8000S, production model, 1984

117

Nokia Cityman

1987

With cellular car-phone manufacturer, Mobira, Nokia had been making car phones and the phone-in-a-suitcase Mobira Talkman, for several years before it brought out its first hand-portable phone. Another collaboration with Mobira, the Cityman was a monster of a machine that weighed 790 g (1 lb, 11 oz), although subsequent variations were smaller.

But that wasn't the point. The point was that it was a marvellous feat of engineering, a fifth of the size of the Mobira Talkman, and that it was 'a complete telephone in your hands'.

Ignoring the fact that the only reason any sensible person would carry it around would be to knock out unsuspecting muggers, the very expensive Cityman was squarely targeted at the business market. Available in black, grey or cobalt blue, it was an impressive phone, and was serious competition to Motorola's DynaTAC, the only other hand-held phone on the market. You could store all your numbers in it, control the volume, and it even had a four-level ringing tone that could easily be heard, even from the leather briefcase it was supplied with. Incoming callers could leave messages, though what we know today as voicemail was then a secretarial service.

The Cityman was a phone, said Nokia, for power-people who needed a direct personal telephone link everywhere they went, and who

The original Nokia Cityman,

couldn't afford not to communicate their 'messages and decisions' immediately.

The marketing material attempted to persuade potential customers of the value of mobile communications by urging them to think of a situation where they could not imagine making a telephone call 'then take the Mobira Cityman with you!' Skiing for example: 'You can receive a call even when coming downhill.' Or jogging, perhaps – carrying the Cityman would certainly help keep you fit.

Nokia Cityman 5000, a smaller, lighter model from the early 1990s

Nokia 2110

The Nokia 2110 was – if only for the interface – the mother of all phones.

In 1994, following a nasty recession, there still weren't many phones on the market, and the limited choice on offer was expensive and bulky. When Nokia launched the 2110 it became an instant classic. It was the design that finally took phones to the people. The 2110 simplified the layout of alphanumeric keys into four rows, located function keys above them in a pretty flower arrangement and had a display that was enormous for its time.

Smaller than a bulky wallet, comparatively lightweight and ergonomic, the 2110 was compatible with most laptops and palmtop organizers, and versatile enough to allow such high-tech functions as sending faxes and emails as well as plain old voice calls.

By 1995 Nokia's global phone sales had shot up dramatically. By 1996 phones were worth more to

Nokia than any other part of its business, which at the time ranged from tyres to wires.

The 2110 became an icon very quickly, and its interface might as well have been an industry standard for the amount that it influenced subsequent phones by Nokia and everyone else. It was widely credited for Nokia's growth into the long-standing market leader, and both phone and company won numerous awards.

Motorola StarTAC

What is lightweight, fits perfectly in the pocket of your Levi's, is roughly the size of a pager and lets you order dinner on the run? It's the Motorola StarTAC wearable cellular telephone!

Motorola was leading the way with smaller and smaller models and extreme portability, and the StarTAC was the smallest phone yet. Motorola marketed it as a 'ready-to-wear' accessory, which could be clipped to a belt in a holster or worn around the neck.

The StarTAC's design was clearly, unashamedly, taken from its futuristic predecessor, the Star Trek Communicator. Perhaps Motorola, which had spent much of the last couple of decades making radios to use on the moon, was projecting its hopes for Neil Armstrong onto its customers with a phone that looked like it would work in other galaxies.

As the first earthly clamshell phone, the StarTAC won awards in Japan and Europe. Its flip-up cover was a small step for phones, but a giant leap for convenience, creating a new category of phones as fashion. The StarTAC meant freedom: its size meant that this phone could boldly go where no phone had gone before.

'The StarTAC phone is not just small, it is small in the right ways,' read one distributor's press release. 'When opened to its full size, the body-friendly StarTAC phone comfortably forms to the face to maintain the familiar ear-to-mouth ratio for which Motorola's phones are known. When folded, the StarTAC phone is so small and light it can be worn fashionably as an accessory. This wearable wonder stands apart from the wide range of portable and personal phones available today.'

The StarTAC was the first cellular phone able to operate with two removable batteries at the same time, meaning users could have up to four hours of continuous talk time or up to an incredible 47 hours of standby time. Other features included a 'smart' button, which simplified one-handed use of the phone, and the vibration alert feature to receive incoming calls less obtrusively.

Nokia 6110

The 6110 candybar phone was billed by Nokia as 'the perfect package', as the company's smallest phone yet. Available in blue, green, black or a range of (then) stylish shimmer effects, Nokia's famous easy-to-use interface was developing and making a strong impression with consumers.

In 1996 this phone included everything anyone knew they wanted. It felt solid and comfortable to hold, it had good-quality sound and a large display. It was the first phone that let people choose the calls they wanted to answer. Those who can't imagine a time without being able to divert the boss/mother-in-law/landlord to voicemail, thank Nokia.

The versatile 6110 was also the first phone to feature games. A major selling point, users could play Snake, Memory or Logic when bored or lonely or, even better, they could challenge other 6100 owners using the built-in infra-red link.

Nokia 8110

Sophisticated at the time, and with amazingly retro good looks, the Nokia 8110 (a.k.a. the banana phone) became a best seller. Commercially available in Singapore from 13 September 1996, it was a high-end, lightweight phone with an elegant design and enhanced quality and functionality.

Looking at the ergonomic requirements of phones from an entirely new angle gave Nokia a model that also felt good to hold. The unique curved handset was designed to fit the face's contours, and a sliding mouthpiece aimed 'to maximize voice clarity'. It also kept the 8110 small enough to fit in any large pocket, while protecting the keypad when closed, preventing calls being made accidentally, a hazard of some other phones at the time.

The 8110 was the first Nokia phone to feature a dot-matrix full graphic display, which changed the text size for easier viewing. Further interface fun could be had as the Nokia 8110 was the first phone to offer Asian language options, including Chinese, Bahasa Indonesia, Bahasa Malaysia, Spanish (for the Philippines) and Thai.

Nokia 8810

According to the Nokia 8810's press release, this chrome-look handset would really boost your style credentials. The 'epitome of understated cool', uniting stylish design with sophisticated technology, was one sexy phone.

Its bright metallic sheen (described by Nokia as 'hot metal') gave it a contemporary look that could compete with a classy fountain pen or a flash watch for show-off potential.

Taking the slide cover design of the best-selling Nokia 8110 'banana' phone, but reducing it to a feather-weight 98 g (3½ oz), Nokia added a glossy casing and raised the price. The statement was purely visual but in the 8810 Nokia had invented a new niche: an 'exquisite accessory' premium phone.

Nokia 8850

Nokia gave us lots of phones in 1999, but it really only needed to give us four. Between the 3210, the 8850, the 8210 and the Communicator, there was a phone for everyone. And by the end of the twentieth century it seemed like everyone who owned a phone owned a Nokia.

The success of the 8810 in 1998 had shown Nokia that there was a big market for the phone as status symbol. The following year the company took the idea to a larger audience with the 8850, a phone that blasted its predecessor out of the picture entirely.

The most visible difference between the two phones was the cover. While the 8810 looked like expensive shiny chrome, it was actually cheap coated plastic, and after a week or so bouncing round the bottom of a bag, it showed. The 8850, however, was the real deal: matt aluminium finish that was hard to scratch, so the phone kept its good looks for longer.

Nokia updated the software, included the new high-tech predictive text facility and improved the battery. If you had the money, it was widely accepted that the 8850 was the phone to spend it on.

Nokia 9000 Communicator

The pocket-sized Nokia Communicator was a huge success, combining all the functions needed in a mini office for when you were on the move. A compact unit the size of a glasses case opened up clamshell-style to reveal a full keyboard and a large, backlit display screen.

Besides all the amazing new functions – email, internet, image viewing, diary, notepad and calculator – it was the QWERTY keyboard with decent-sized keys and a large screen that made it special. Writing a fair amount of text on it was a real option.

Business people snapped up the phone for its sheer usefulness, but, more surprisingly for Nokia, the Communicator was also bought in huge quantities by minorities that are all too often excluded by design. It was first phone to offer visually impaired users affordable access to SMS via text-to-speech, though the software had to be bought separately from a third party.

Meanwhile, deaf users also found the text facilities remarkably convenient. The Nokia Communicator was the ultimate communications tool of its time.

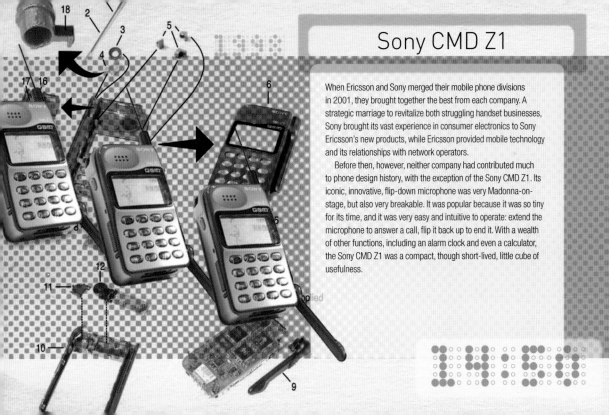

Sony CMD Z1

When Ericsson and Sony merged their mobile phone divisions in 2001, they brought together the best from each company. A strategic marriage to revitalize both struggling handset businesses, Sony brought its vast experience in consumer electronics to Sony Ericsson's new products, while Ericsson provided mobile technology and its relationships with network operators.

Before then, however, neither company had contributed much to phone design history, with the exception of the Sony CMD Z1. Its iconic, innovative, flip-down microphone was very Madonna-on-stage, but also very breakable. It was popular because it was so tiny for its time, and it was very easy and intuitive to operate: extend the microphone to answer a call, flip it back up to end it. With a wealth of other functions, including an alarm clock and even a calculator, the Sony CMD Z1 was a compact, though short-lived, little cube of usefulness.

Nokia 3210

Nokia's cheap-and-cheerful 3210 was designed to lure first-time buyers into the wonderful world of phones by offering colour, fun and freedom of choice at an affordable price.

The 3210 brought about a mini revolution with its Xpress-on changeable covers, cashing in on the trend at the time for all things customizable and bringing the concept of personalization to consumer electronics. Users could completely alter their phone's appearance by dressing it in a 'funky zircon green' or 'cyber fish' cover, and they loved it.

Another new feature for the 3210 was picture messaging. A million miles from photo messaging, it was usually basic clip art with titles such as 'flying heart', 'birthday cake' or 'dancing' that could be used to cheer up – or even thoroughly depress – your friends and family. Although the feature would only work to and from another Nokia 3210 phone, that wasn't a problem because everyone seemed to have one.

Nokia's new phone was also one of the first to incorporate predictive text. 'It knows what you're going to write before you've even finished!' exclaimed the press release. It successfully baffled everyone for months.

Nokia 3310

A revamp of the 3210, the 3310 was just as important in the history of phone design, simply because so many people had it. The 3310 offered something new and different, something that no other manufacturer had succeeded in designing: a solid, basic phone.

The 3310 looked better than its predecessor, being smoother and rounder. Functionality wise, it was a smoother all-rounder too, with all Nokia's latest features incorporated, such as a 'chat' mode for text messaging and new games, including Space Impact and Bantumi.

Teenagers loved it because it was cheap, cheerful and had Xpress-on covers. Builders, busy mothers and clumsy people loved it because it was one of the toughest standard phones around. If you dropped it, it bounced. First-time phone owners loved it because everyone else had one and it was incredibly easy to use.

The unfussy, trustworthy 3310 was not just a popular phone, it was a democratic one. With it, Nokia took technology to the masses.

Nokia 8210

The 8210 was Nokia's smallest, lightest, skinniest phone to go on the market. But it wasn't just adorable, it was also impossibly cool.

For any consumer electronics at the time, especially phones, small was best, but Nokia was careful not to compromise any features in order to get the 8210 down to size. Weighing just 79 g (3 oz) it could do everything all other Nokia phones could do – predictive text, voice dialling and picture messaging. There were Xpress-on covers and thirty-five ringtones to choose from to express your mood. Yes, the 8210 could do all these things. And it could do them while being very, very small.

Of course, not everyone was happy with Nokia's latest offering – its competitors, for example, when the 8210 started flying off the shelves. But the mini-mobile was also responsible for a new debate in the industry: could a phone be too small? The question had never been considered before – clearly, the smaller the technology is the less we notice it and the more sophisticated we all are, right?

But the 8210 made us realize that this idea was faintly ridiculous, partly because most people's mouths are more than three inches away from their ears, and there were complaints from fat-fingered text addicts.

Global opinion was divided: the Japanese loved it, Americans hated it and Europe was not quite sure. But the profits rolled in, and Nokia, once again, had built an icon.

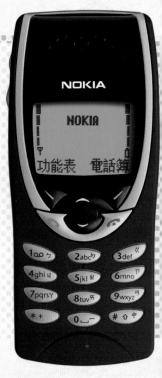

Nokia 7110

When *The Matrix* came to our screens its huge success was thanks to several factors, but above all it was because it made the future look like a damn cool place to be. The props were fantastic, and everyone wanted one of the *Matrix* phones, which had been dreamt up by Nokia for the occasion.

Nokia were feeling futuristic and recklessly predicted that by 2003 more people would be using phones to access the internet than PCs. They heralded this internet revolution with the launch of the Nokia 7110 – the first WAP phone, designed to bring the internet to the 'man in the street'. Of course WAP eventually flopped, but that was nothing to do with the phone. The original *Matrix* phone

(Samsung also produced one for the second film in the trilogy; see p. 166), the 7110, went into production with an enlarged screen for viewing emails and web pages, and fans went wild. It had an ergonomic, curved shape and a sliding cover over the keypad. It also introduced the ugly but useful NaviRoller key, positioned in the middle of the phone and controlled by the thumb to scroll through menu options at speed.

IDEO Kiss Communicator

IDEO, a global product-design consultancy, has a particular, peculiar way of thinking about technology, which has often seen them hit on revolutionary new ideas. While manufacturers were telling the world what all their new technology could achieve, IDEO asked what people might, in fact, want it to do. One result produced by its London office in 1999 was the Kiss Communicator. Here was a real innovation: an electronic device that had the power to actually delight people rather than slightly frighten them. The designers had noticed that pagers were often sold in pairs to couples who would send romantic messages to one another during the day.

With the Kiss Communicator lovers would also buy a matched set. If one person blew a kiss into the device, it would be picked up by sensors and transformed into an abstract pattern of light, which would then be transmitted to his or her partner's Communicator, who could then answer in turn by squeezing the glowing product's sides. A pattern of lights is then displayed back to the sender.

The Communicator was the first mobile device that was designed to transmit something more fundamental and emotional than text, voices or images: this was a virtual kiss, a squeeze and a smile....

RIM BlackBerry

Blackberry – until 1999 it was just a tasty fruit that was good in jam, after then it was also a useful device that enabled you to receive your email when away from the office. By 2000 the BlackBerry had evolved into a much bigger, very useful device that was also a phone.

The BlackBerry 'pushed' email to the phone in the same way that we were by then used to receiving text messages. No dial-up was required. From the start, the BlackBerry email phones gathered a loyal following with workaholics. Its no-frills design was targeted at the corporate market and it proved especially popular in the United States where users quickly nicknamed it 'CrackBerry' for being so addictive.

The BlackBerry also became something of a social problem. Anthropologists were already fond of discussing how our state of perpetual contact was affecting our social lives. The BlackBerry was something else.

Incorporating a small yet usable, twenty-six-button QWERTY thumb keypad (which became widely imitated), faster and easier typing was possible than with the typical ten-button keyboards of conventional handsets, while the screen size was about thirty per cent larger than typical phones, fitting up to twenty lines of text. And once email became portable, people could be available any time and almost anywhere. The BlackBerry made it easier and more tempting than ever to work during personal time and sneakily catch up with friends while in the office.

RIM BlackBerry 7100T, 2004

RIM BlackBerry 7100V, 2004

The intrusion didn't come cheaply, however, and BlackBerries were mostly used by highly paid professionals. For some the device was a symbol of importance: you knew you were indispensable when the company provided you with one of these, and that has a certain cache.

above RIM BlackBerry 950, 1999
left RIM BlackBerry 7200, 2003

Handspring Treo

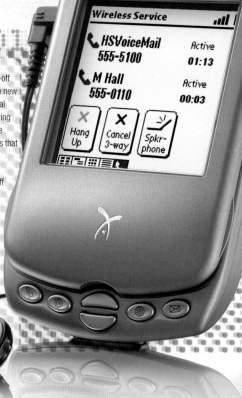

Worn out from carrying three gadgets around to show-off to your friends? You might have been at the start of the new millennium, what with your pager, phone, flash personal organizer or any combination thereof. Luckily, Handspring provided the answer with its Treo range of three-in-one 'communicators' – one of the first combination devices that would make it big.

'The idea for Treo was to combine a phone, a pager and an organizer into one small product', explained Jeff Hawkins, founder of Handspring and chief architect of the Treo family. The first two Treo products, he claimed, would mark the beginning of a new era of innovation and 'will fundamentally change the way people stay in touch.'

The Treo worked like the RIM BlackBerry but without the always-on connection. With organizational capability as a bonus, the Handspring Treo was primarily built to communicate.

It featured an elegant interface, easy browsing
the internet and short messaging and email
actions in one handy package.

The design, clearly influenced by the Motorola
clamshells, and perhaps there's a little Star Trek
communicator in there too, had a transparent
cover that flipped up to answer calls before
acting as the earpiece for the phone, while a
microphone was located at the bottom of each
unit. One model had a handwriting recognition
function for those who might find the tiny
keyboard keys too small for comfort.

left Treo 650, 2004
t Treo 270, 2002
ght Treo 650, 2004
right Treo 90, 2002

Vertu Signature

Vertu was a spin-off of Nokia created to fill the gap in the market for über-expensive, luxury phones. Understandably, as phones were proving so disposable, constantly being rendered defunct by new technology, nobody had thought of this before.

Vertu created phones in another league: if Sony Ericsson was Audi, Samsung was Toyota, then Vertu, without question, would become the TVR of mobile communication. The Signature model contained 'enough platinum for twenty wedding rings', though it was also available in eighteen-carat gold. It was for rich people who really wanted the world to know they appreciate quality. It was pure flash.

Vertu was the brainchild of Nokia's creative director, Frank Nuovo. It was a success because they didn't need to sell many phones at $10,000 to make a profit, but they did anyway because some people will always spend money if they think they are buying the best. And in the luxury category there was no competition.

Assembled by hand in Vertu's workshop in Hampshire, UK, Vertu's internal architecture featured more than 400 mechanical parts. The 'perfect click' was achieved by hand threading a microscopic rubber band onto a tiny pin, and then placing it between eighteen jewelled

bearings underneath the keypad. Yes – jewelled bearings. Underneath the keypad.

Vertu's phone boasted a hi-fidelity audio system speaker with original composition ringtones. Using 'spacecraft technology', a pillow for the earpiece was created that was warm to the touch and durable. To ensure the logo wouldn't wear off, it was applied using physical vapour deposition in a vacuum chamber. Likewise, not satisfied with glass for the display (presumably that would be too ordinary), Vertu developed 'the largest sapphire crystals ever commercially produced' and cut it with diamond cutters.

Best of all, Vertu phones all featured a dedicated concierge key on the side. This magic button allowed clients to connect to a team of operators who would book your theatre tickets or spa appointments, on call twenty-four hours a day. If nothing else, the Vertu phone would bring a whole new meaning to the term 'upgrade'.

IDEO Social Mobiles

Manufacturers are generally fantastic at spending all their money on finding the next killer application for phone technology. They are extremely good at devising services we don't yet know we need and producing smaller, lighter, louder, uglier and generally more ridiculous all-singing, all-dancing phones to cram them into.

One area that has been seriously overlooked in this frenzy of research, however, is the social aspect of phone use. The Social Mobiles project by Graham Pullin and a team of designers at global product-design consultancy IDEO and design theorist Crispin Jones explored just this, and uncovered five, quite different potential new services.

Phone etiquette is generally not considered to be the manufacturers' problem. It's up to their customers if they want to make fools of themselves by insisting on a tacky ringtone, or shouting loudly into it while on public transport. But the more phones become integrated into society, the more ways their use can offend. Could these problems be addressed by the design of the phones themselves?

The Social Mobiles project concept phones looked at a different common social difficulty. The results were phenomenal, and the phones were featured in *The Economist*, *Wired*, *The Independent*, on the radio and on television. The project also won a Media Art

in Tokyo, where all five ideas
received especially well.
Mo1 was intended to be
tory for 'repeat offenders'
uld be given to those people
ersistently disturb others with
ve phone calls. SoMo1 teaches
er to be more considerate by
g a variable electric shock
ding on how loudly the person
other end is speaking. As a
both parties are induced to
more quietly.

Mo2, for those who already
to be considerate but may
take a call urgently, allows
converse silently. A person
ng a call in a quiet place, such
art gallery, can respond with
but expressive vowel sounds
that they produce on the handset's
touch-sensitive dial. It is the
antithesis of text messaging in that it
conveys rich emotional nuance at the
expense of textual information. And
considering how much conversation
consists of ooohs, aaahs and mmms,
it could just work.

Perhaps the most comical,
SoMo3, is half phone, half toy
trumpet, and requires the user to
play a tune for each phone number
they wish to call. Turning dialling into
a public performance will make them
think twice about how appropriate it
might be to make a call.

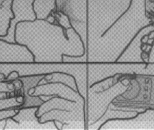

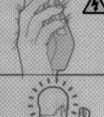

Siemens Xelibri

Xelibri 5

The Siemens Xelibri line, launched to great fanfare in 2003, was certainly influential, but for all the wrong reasons, notably that it was such an embarrassing flop. A radical step in the mobile-phone market, Xelibri were more than phones as fashion accessories: they were fashion accessories that made phone calls.

They did precious little else, and they did that badly. The bizarre 'space on earth' keypads (some didn't even have that) were hard to master, and the designers had no idea what the word fashion meant. As the press release would have it, something could be made fashionable by fitting three criteria: the phones would be wearable; they would be sold in clothes shops and department stores; and the designs would be driven by form, rather than function – cue lots of quirky shapes, bright colours and 'individuality'.

Xelibri 8

Observing how many people were attracted to good-looking phones, and that they would often show them off, Siemens saw the potential for them to become as much of a fashion accessory as watches, handbags and shoes. 'We envisage the scenario where people will own many fashion accessory phones and wear the one that matches

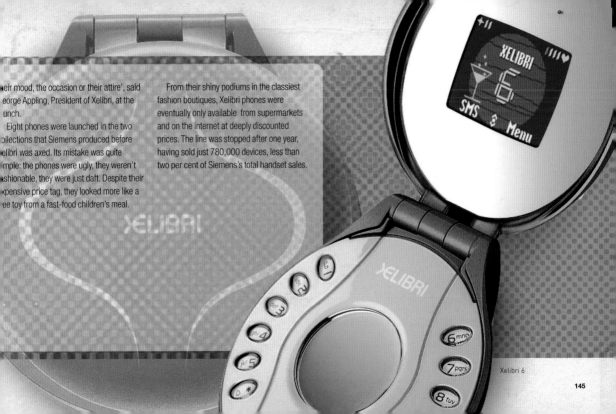

eir mood, the occasion or their attire', said eorge Appling, President of Xelibri, at the unch.

Eight phones were launched in the two ollections that Siemens produced before elibri was axed. Its mistake was quite mple: the phones were ugly, they weren't shionable, they were just daft. Despite their xpensive price tag, they looked more like a ee toy from a fast-food children's meal.

From their shiny podiums in the classiest fashion boutiques, Xelibri phones were eventually only available from supermarkets and on the internet at deeply discounted prices. The line was stopped after one year, having sold just 780,000 devices, less than two per cent of Siemens's total handset sales.

Xelibri 6

Motorola V70

When it was launched, Motorola claimed its new phone was nothing less than a 'Motomarvel'. Indeed, the design-conscious V70 quickly became the phone in all the most fashionable circles, not least for its sleek and shiny never-before-seen rotating cover and circular display.

Motorola had replaced the flip cover it was known for with one that rotated a full 360 degrees around the round screen. It was a stunning, jaw-dropping makeover that exuded style and sophistication.

Unusually for a style phone, it was also comfortable to hold and extremely light. But perfection is elusive, and sadly the V70 was let down by Motorola's interface. At a time when people had got used to the ease of use that Nokia could offer, nothing else would do. The display was also minimal, with only room for two lines of text at once. Its rocketing success was only matched by its fall from grace – fashion is fickle, and rarely more so than in this market.

Regardless, the V70 was gorgeous while it lasted, and it was completely different to anything that had gone before. Fully customizable, the circular silver bezel that framed the menu could be replaced with one in gold, silver or white. It was expensive, and therefore a status symbol and, for those who really wanted to show off, it could be worn around the neck using a (Gucci) lanyard. Pure class.

Samsung T100

A little fashion phone of fun, the Samsung T100 was the first phone available in Europe to have a colour screen.

The T100 wasn't a camera phone, as camera functions weren't a given then, so the colour was mainly to make the menus look good. You could connect the phone to a PC and transfer pictures to use as the background too, which was still a novelty you could impress your friends with then. There was also a secondary dual colour display on the front of the phone for when the clamshell was shut. The two screen phone was one of Samsung's trademarks, and as the outside screen showed what was on the inside screen, users could see who was calling when the phone was shut, and therefore avoid them. A call-screening screen, you could say.

The T100 was a simple, pretty and popular phone. It was probably one of the first to be designed specifically for the female market – being curvy, light and super-polyphonic. It was never meant for gadget lovers. Despite being able to connect the phone to a computer, it wasn't possible to synchronize your appointments with it or copy names and numbers over, and it predated Bluetooth. But it was one of the tidiest phones on the market at the time, and it sold.

Sony Ericsson T300

Japan and Sweden are both countries that traditionally excel when it comes to design, so when Sony and Ericsson got together it was a match made in heaven. The two companies immediately spawned a whole range of beautiful new handsets, including one in particular, the T300.

What was unique about it was its shape. Proving a phone can be a brick without being bulky, the T300 had a rectangular prism-like form, while its subtly rounded corners gave it a sophisticated edge.

Similar to its older sister phone, the T68i, they shared the same clean, distinct lines. It was refreshing to get away from complex surfaces, and the phone's stripped-down appeal drew people's attention to the exciting technology underneath, large colour display and recessed navigation joystick.

KDDI Infobar

Japanese genius Naoto Fukasawa designed the Infobar as a project for KDDI's au concept programme [see p. 222]. The design world was stunned but very happy when the company was brave enough to put it into production, only to be disappointed again with the news that it would, of course, only be available in Japan.

The Japanese market took very well to Fukasawa's vision, demonstrating to manufacturers that retro worked, that multicoloured keys were a good idea and that simple can also be sensational.

Fukasawa had originally designed the coloured square keys to be interchangeable, for maximum personalization and fashion potential, but in the production version KDDI fixed them down. The phone was still a kinetic masterpiece and completely different to anything else available.

Several versions were produced, each with unmistakable Japanese references. Ichimatsu's black-and-white checks were based on a classic kimono pattern; Nishikigoi emulated a Japanese koi pond – the blue water alive with splashes of fishy red; Building conjured-up images of a Tokyo skyline, with silver

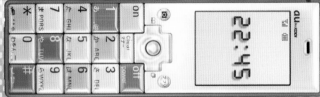

tiles against a night sky; and the delicious Annin is quite clearly Annin Tofu almond jelly.

Not just a pretty candybar, the phone was equipped with high-performance functions, including video and still cameras, to compete with phones at the top of the industry, all contained in a very compact, very sleek magnesium alloy body.

Siemens SL55/65

SIEMENS

SL 65

A jewel. A pebble. Half an egg. There was no end to the poetry that the Siemens SL55 inspired when the reviews came out.

Designed by European agency Designafairs, the ergonomic SL55 was just 8 cm (3 in) long and fitted into the palm of the hand.

Just as appealing, and another design first, was the immensely satisfying switchblade-style sliding cover. It was also as multifunctional and efficient as it needed to be.

The SL55 originally came in a choice of two colours: ruby red and 'titan' (Siemens described it as the colour of black diamonds, but dark grey might be easier to visualize). The company subsequently produced the phone in various limited-edition guises. At last – remembering its recent failures with the Xelibri range of fashion disasters [p. 144] – Siemens successfully played the phone-as-accessory card with its Escada model, a collaboration with the fashion label.

The fuschia pink phone came with a silk cord to transform it into a wearable accessory, and the collaboration marked the start of a happy relationship between the two companies.

OrangeSPV

In late 2002, Orange launched the SPV, the first commercially available handset running Microsoft's SmartPhone operating system. A few months later it surpassed itself with the SPV E100. (SPV was the name invented by Orange for all their Windows Mobile powered phones and simply means 'sound photos video'.)

The E100 was better looking than the original, boasting a larger keypad and a small rocker control. Not long after this, the E200 broke new ground in the smartphone field again, this time including an integrated camera and Bluetooth.

It was only when the C500 came out, however, that people really took notice of Orange's SPV phones.

The idea was to create a mini-computer in the form of a pocket-sized phone. The C500 was the first in the line that was not only a decent mini-computer but also a good phone, and all squashed

into a compact handset that weighed under 100 g (3½ oz). It was, said Orange, a 'marvel of miniaturization'. As well as being smaller, the operating system had evolved to be much more pleasurable to work with.

Because smartphones lacked the touch-sensitive screen found on hand-helds, the interface needed to be driven entirely by the keypad. In the C500 this worked well, with six control keys above the number pad, including a rocker control that, for once, wasn't dependent on the user having skinny fingers. Another nice detail was a light sensor below the keypad so that the keys would only be backlit in low light situations.

Nextel Push to Talk

For a long while the United States's mobile-telecoms industry was fragmented at best, horrendous at worst. Coverage was poor, billing was complicated and cell phone use was accordingly low. Text messaging, an astounding phenomenon throughout the rest of the world, didn't initially catch on.

But then came Nextel with a completely new network technology: push-to-talk (PTT). Nextel subscribers could talk to each other instantly by holding down a button and speaking into the phone, essentially turning a phone into a long-distance walkie-talkie.

Officially released in 1996, the technology was called iDEN and made for Nextel by Motorola. It was instantly adopted by builders, maintenance technicians and safety workers for talking across short distances.

It wasn't a particularly cool or progressive application for network technology, and many people didn't recognize its usefulness for years. But nearly a decade later, with a little innovation, some clever marketing and a healthy sense of competition, Nextel launched its international direct connect service in 2004.

Subscribers in the United States loved it. They could get through to somebody straight away, avoiding the expensive annoyances of voicemail, engaged signals and missed messages. They quickly became dependent on PTT, and it was tipped as their answer to texting. And perhaps it could be useful elsewhere too.

Motorola 3G

The first 3G offering from Motorola had the same name as a jumbo jet, and its size was also fairly jumbo, but the A380 was one of the few choices if you wanted to subscribe to the much-hyped 3G networks in 2003.

Weighing 210 g (7½ oz), and bulkier than the competition from NEC [opposite], its camera was an add-on that was bigger than the phone itself. Surprisingly, Motorola hadn't included the video-calling function. Although not enough people had 3G phones to make video calling a viable option, it was still the main selling point of the expensive 3G handsets. Otherwise – as many people were still asking about 3G in general – what was the point?

But the A380 did everything else, including playing MP3s. It also had a bigger, better screen than NEC's e606, though with far fewer colours, making it ideal for watching video clips and pictures, at least until the battery ran out (fast). Welcome to the world of 3G: most people decided to wait.

Better things came to those who did. The A1000, said its press release when it was launched in 2004, was 'a smart phone for a smarter you'. Lighter and more streamlined, and this time with a very capable video camera function, this was where 3G really began to look interesting.

Finding Nemo cont
to stay in the top 1
after 8 weeks.

5 Unread Emails
Soo Jin
Picture Messages
Thom Rahe
Appointments
Tasks

10:00

NEC 3G

The NEC e606 was one of first 3G phones available and, as such, it suffered when the network had teething problems and failed to live up to its claims. Hundreds of customers were left with poor service and strict contracts, and as a result reviews were hostile.

With 3G, for the first time in phone history, handsets got bigger. The e606 was a brick, like all the early 3G phones, but it had to be to include all its super-functionality. With an integrated digital video camera, 3G operators heavily promoted features such as video calling and downloading and playing video and audio clips. A big display (65,000 colours) was crucial.

The hefty weight and poor battery-life seriously reduced the usability of the first 3G phones. But it was a milestone, and the e606 introduced the wonders of 3G to many.

In 2004, however, the company announced the release of the N900i in Japan, which, with its stylish (ergonomic even), arc-shaped side profile, offered three-times longer standby time and weighed twenty per cent less.

T-Mobile Sidekick

It was a PDA. It was a phone. It was a data communicator. And it had a really sexy swivel screen. It was the eagerly anticipated 'Danger Hiptop', as it was known during development, by Danger Research for T-Mobile Wireless.

A rare example of a truly innovative product design, the Sidekick was an experiment with scale – much smaller than a laptop but still bigger than a PDA. Users were expected, as the witty 'hiptop' tag suggested, to wear the Sidekick in a holster on their hip.

When closed there were three function buttons and a multicoloured scroll wheel. Rotate the screen and you have a decent keyboard with much more room to manoeuvre than a BlackBerry or a Handspring Treo. There was the added bonus of a directional pad that was extremely handy for, among other things, playing games. The Sidekick never had

pretensions to be a business product, so didn't step too hard on the toes of its closest rivals. Danger Research wanted to capture the youth market, and the rock-solid data device it had produced looked great and lived up to its billing.

Like most combination PDA / phone devices, however, the Sidekick presented a compromise. As a PDA it was reasonable, but as a phone it could be awkward and required users to plug in a headset because of its bulk and weight. The object might have looked cool, but it was pretty hard to do the same if you were using it.

Sharp Vodafone live!

He looks, he shoots, he scores! When Sharp and Vodafone joined forces to make a camera phone with which to launch the new Vodafone live! service, they paid David Beckham to endorse it in the UK, Spain and Japan. The footballer was a male style icon, and he duly did for phones what he had unwittingly done for sarongs before.

The GX10 (along with subsequent revisions), was available exclusively through Vodafone and quickly became known as the Beckham phone. A clamshell phone with an advanced camera for the time, the handset was a vehicle for the Vodafone live! services: rich-media games, far too many polyphonic ringtones and enough web content to satisfy most mobile needs of the time.

Design-wise the Beckham phone is a slightly more masculine version of the Samsung T100 [p. 148], with an external screen as well as a bigger flip-up display on a satisfyingly chunky silver-grey block. A mark of its build quality was that second- and third-generation designs, which incorporated new services and features such as a Vodafone live! dedicated button, changed very little physically, bar getting a little fatter.

Sony Ericsson P800

Wildly anticipated before its launch, Sony Ericsson's P800, was a hard-hitting gadget, the ultimate mobile multi-media experience at the time. This is despite its relatively unsophisticated appearance in pale blue and silver-grey cheap-looking plastic. According to Sony Ericsson's president, Katsumi Ihara, it was a phone that would 'change the way people communicate' and it would 'help people create new ways of expression and interaction.' And if you thought that was the whole purpose of phones anyway, this one took Ihara's words to a new level.

The P800 was the first palm-sized device to use the extremely user-friendly Symbian operating system. It was also the first to sport a touch-screen. Perhaps radically of all, it wasn't made by Nok

Incredibly small when you conside the functionality it harboured, the P80 comfortable to hold and use in one ha and it made the operation of two diffe devices – PDA and phone – with one display seem seamless.

When closed a flip hid three-fifths the display, and you could do almost everything you needed by means of t jog dial and buttons on the flip. Open flip and the touch-screen was activate for a PDA-style system. What was cle was that the flip itself contained no electronics, and instead used plastic plungers that pressed corresponding points on the touch-screen undernea it when closed. Although it did have it

SONY

MEMORY STICK

Memory Stick Duo Adaptor

uses, the flip could be removed for those who didn't see the point.

Best of all, though, the P800 came with four flimsy looking plastic styluses – one to use and three to lose – to prevent getting grubby fingerprints all over the touch-screen.

Sony Ericsson T610

Erik Ahlgren, the designer of Sony Ericsson'
popular T610 phone, claims to have been
inspired by several things for this commissi
digital cameras, high-quality stereo equipm
and, of course, Sony Ericsson's stylish
company brand values.

But thankfully something more original d
also enter his mind, and as a result the T61
looked completely different from anything
else out there. The glossy black lacquer-lool
finish that made the T610 so covetable was
inspired by a grand piano. The shiny black
was balanced by aluminium panels. In som
countries versions were available with colou
themes based on the elements of nature
('volcanic' red or 'abyss' blue). The only
drawback to the glossy new finish was the
fingerprints it attracted. But with a cleaning
cloth included in the package, no one worrie
too much.

Sony Ericsson was establishing a good reputation for its phones and, even before they had a chance to hold it, technology enthusiasts were praising it as the perfect phone. The company cleverly marketed it as a beautiful accessory for beautiful people and soon – naturally – everyone wanted one.

Despite echoing the brickishness of previous Sony Ericsson models, the T610 didn't have a single straight line bar those surrounding the screen. The curves were so subtle it was easy to miss them but, according to Ahlgren, their presence made the phone infinitely more ergonomic and aesthetic. And the sales figures tended to agree.

Nokia N-Gage

Nokia brought mobility to the games industry in 2003 with N-Gage, a product that had games-console manufacturers quaking in their boots.

Happy to compromise the phone experience for an audience with other applications in mind, N-Gage was a games console that was also a phone, rather than the more traditional approach of a phone that could play a few games.

N-Gage was designed to take high-quality mobile games to a new level by offering them over the networks. As network capacity increased, there was a greater potential for interactive games using Bluetooth and wide area gaming. Gamers could communicate properly with each other and form a larger, more powerful community of fun.

Nokia worked with games publishers and developers to provide the most appealing catalogue of games, which were distributed on memory cards.

Making one of the first steps towards converging trends in technologies and devices, Nokia was playing for the jackpot.

A predictable development maybe, but a phone that was also a television was a welcome one.

NEC released the N940, a phone with a built-in television tuner, in China at the start of 2005. Designed to let users watch up to sixty minutes of analogue television while making and receiving phone calls, it was finally possible to get your soap-opera fix or keep up with current affairs on the go.

Not sacrificing anything if it could possibly help it, NEC also built in a high-quality digital camera – a novel treat when combined with a huge colour display.

For all its high-tech gadgetry, the N940 also incorporated a wonderfully retro feature: a pull-out antenna. It was necessary for television reception and it was endearing.

Given that portable televisions never really took off the first time round, what made NEC sure that people would go for it now? Partly, getting decent reception was at last possible, but also that it was now attached to a phone and a camera.

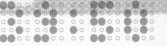

Samsung S250

Like Nokia had done for the first *Matrix* film [p. 133], Samsung designed a handset for the second in the trilogy, *The Matrix Reloaded*. Samsung put the *Matrix* phone, the N270, on sale in a limited edition of 10,000, in time for the film's release on 15 May 2003.

This *Matrix* phone was an uncompromising expression of technology. The type of phone that looked like it would explode if you threw it, like a grenade. But it boasted lots of features, and – crucially – it had an exciting pop-up design: the earpiece popped up to reveal the LCD.

In 2004 Samsung was making waves again, but this time with a really commercial proposition: the first ever five-megapixel (that's proper digital camera

quality) camera phone, the S250.

Exciting in its own right, the S250's unique stretch design was derivative of the *Matrix* phone, but with much, much better styling and materials.

Just one year after its first one-megapixel phone and three months since the first 3.2-mexapixel offering, the five-megapixel S250 was an awe-inspiring achievement for Samsung. With a huge memory, it could also store up to a hundred minutes of video.

left and right Samsung S250, 2004
far left Samsung N270 *Matrix* phone, 2003

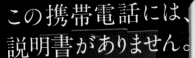

それは、説明書がいらないくらいカンタンだからで

通話専用ケータイ
ツーカー S

THE NEW VALUE FRONTIER
KYOCERA

Kyocera TU-KA S Phone

Wireless carrier TU-KA launched the S Phone for those people who don't want all the bells and whistles of modern gadgetry. People who wanted a phone that was just for making phone calls with.

The S Phone featured big, easy-to-read numbers and…not much else. There was no screen, so of course no camera, no games and no menus. It had no choice of ringtones, it wouldn't email, send or receive text messages, store music, play video or any of the other things that phones generally had as standard, there wasn't even a caller ID feature.

Basically (and that's the operative word) you could do four things with the S Phone: turn it on, turn it off, dial a number and accept a call. And the battery lasted for ages.

It was controversial, partly because it was so outrageously simple and partly because it was so popular. Elderly people loved it for its big buttons and familiarity, while parents bought it for their children.

KDDI talby

When Marc Newson's talby phone hit the shelves in Japan, hip Europeans and Americans turned green with envy, and not for the first time – Japan had been way ahead in the phone market for years. But this was one of the best looking phones by some way, and for some time, and it evidently had global appeal. It didn't actually do anything especially new, so why was this kind of design not available in the West?

The third designer to enjoy the privilege of redesigning a phone from scratch for KDDI, in its prestigious au concept programme [see p. 222], Marc Newson jumped at the chance to do something different. Having put his skills to everything from a private jet to swanky hotels, from expensive watches to expensive trainers, Newson's reputation for designing gorgeous organic forms for an elite, affluent designer clientele preceded him. Designing a phone was an opportunity for him to make an impact on the lives of a wider audience, and a chance to transform an industry that he felt was stale and boring in terms of design.

'When as a consumer I can't go out and get something that I like, I want to do something about it,' said Newson in *Blueprint* magazine. 'I would actually go out and pay money for this – as opposed to the other crap that's out there. Telephones have followed in the footsteps of things like sneakers and cars – the vast majority of everything that is happening is just gross. It's awful.'

Manufacturers seemed to be slowly realizing that, as well

as having extra functions in phones, people were more than happy to pay for something that looked good too. But, as Siemens's Xelibri [p. 144] had already shown, looking cool is not always as easy as it seems. Newson knew how to make fashion function, and as a result everyone wanted a talby phone.

LG 8110

2004

The excitement surrounding 3G networks was somewhat dampened when we saw the size of the phones for them. The LG 8110 was not a sexy phone, but it was the first 3G phone that was not gigantic. If you wanted MP3 ringtones and video calling, it was the first handset that was worth considering.

Besides the size, LG's new handset was like many other phones on the market, which, with all the extra functionality, was a considerable achievement. A clamshell design with a rotatable camera module, the phone could be opened like a laptop, which was useful for video calls, like a normal clamshell for standard voice calls, if you were used to clamshells, and 180 degrees if you prefer candybar phones.

Such user consideration, which is rarer than it should be in phone design, was why the 8110 succeeded where others failed. 3G technology was easier to get a handle on when the handle itself was generic and consistent.

LG KP8400

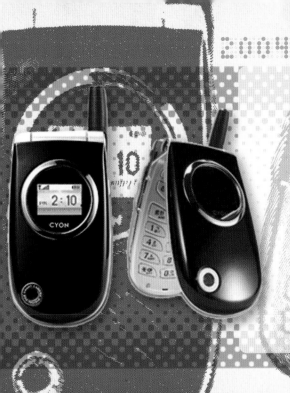

A phone that doubles up as a blood-testing device for diabetics, at first impressions the KP8400 seemed unusual, if not a little eccentric. Think about it twice though, and this marriage of technologies is a straightforward, sensible innovation.

One of Korean manufacturer LG's strengths is in developing phones for markets that other companies consider to be too niche to concentrate on, if they consider them at all. LG developed the Mecca phone [see p. 50] and the self-sterilizing phone – both of which became invaluable to thousands of people.

The blood-testing phone, developed in conjunction with health equipment company Healthpia, was perhaps the model that had the most potential to demonstrate how much leverage there was in such diversification. It included a tester (located near the phone's battery pack) into which users place a strip with a drop of blood on. Insulin and blood readings are then shown on the phone display. They can then be uploaded to an online database for retrieval later on.

The added technology was reasonably complicated in itself, making the phone rather bulky. It didn't incorporate many other flash features, but it was more than adequate with sixty-four polyphonic ringtones and a camera. And if it saved lives, who's going to complain?

Motorola V3 RAZR

Long regarded as an engineering-led manufacturer, in 2004 Motorola conceded that there might be a place for design too. They had spotted a gap in the global market for a phone that would combine fashion and function, and their product development team in the United States was subsequently briefed. A few months later the V3 (or RAZR) was launched.

The RAZR brought a new meaning to the term cutting-edge: an ultra-slim aluminium clamshell phone, Motorola had whittled its width down to just 13.9 mm (approx. ½ in). Its design team boasted that the V3 could hide behind a tower of SIM cards if set at the right angle, and its skinny frame inspired British *Vogue* to call it the 'superwaif of the phone world'.

Not just a pretty face, the V3's good looks and minimalist form disguised rugged engineering. Made from anodized aircraft-grade aluminium, the V3 was as tough as it was light, while an innovative keyboard design was constructed from a single sheet of nickel-plated copper alloy, with numbers and characters chemically etched into its surface. Achieving significant weight loss in a handset can require taking extreme measures. In the case of the V3, so as not to compromise on the

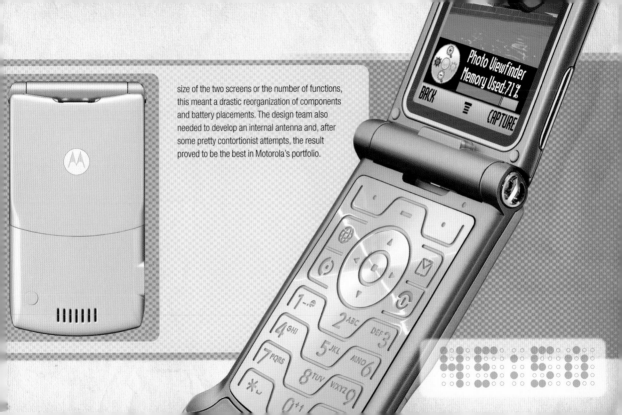

size of the two screens or the number of functions, this meant a drastic reorganization of components and battery placements. The design team also needed to develop an internal antenna and, after some pretty contortionist attempts, the result proved to be the best in Motorola's portfolio.

Photo Viewfinder
Memory Used: 71 %

BACK

CAPTURE

Nokia 7280

At the start of the new millennium Nokia, having led the way throughout the 1990s for producing phones with simple good looks and clear interfaces for the masses, began to experiment with phones as fashion accessories. To begin with it didn't look like a wise decision, and its insistence on pursuing a new design language that favoured form over function wasn't initially popular. Nokia's designers began to change keypad configurations and introduced complicated functionality. The new approach was highly segmented, fashion-based and frivolous.

So Nokia didn't design any especially influential phones for a while. Until 2004, when fashion phones began to come into, well, fashion. Taking a straightforward approach, Nokia simply copied the catwalk trends of the previous season and launched three phones that finally looked like they might push the boundaries of traditional

phone design. The Nokia 7280, Nokia 7270 and Nokia 7260 took their inspiration from the lavish and decadent 1920s. 'The designs', read the press release, 'blend old-world Art Deco styling with an edgy, modern-day twist'. Core style influences, such as flow and movement, colour, geometry, detailing and graphics, were the inspiration for the collection.

There's nothing particularly revolutionary about a Deco-look gadget, but one phone in the trio was far more innovative. The Nokia 7280 offered a completely new phone design, foregoing the traditional keypad for a discreet keyless dial.

With the lacquer-inspired, high-gloss finish, the leather and mirror accents, the 7280 was a significant move away from anything else that was happening in phone design. The long lipstick-sized oblong looked more like a spy

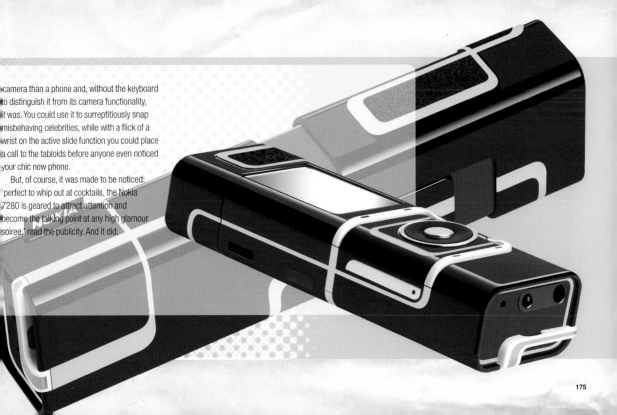

camera than a phone and, without the keyboard to distinguish it from its camera functionality, it was. You could use it to surreptitiously snap misbehaving celebrities, while with a flick of a wrist on the active slide function you could place a call to the tabloids before anyone even noticed your chic new phone.

But, of course, it was made to be noticed: 'perfect to whip out at cocktails, the Nokia 7280 is geared to attract attention and become the talking point at any high glamour soiree,' read the publicity. And it did.

Isamu Sanada HiPod

What if Apple were to make phones? In a market where most products were black and chrome for far too long, and although great design is rife, genuine innovation on a really dramatic, world-changing scale is rare, the prospect is a fairly appealing one. Surely the company whose name has become synonymous with such innovation – iMac, iPod, etc. – could treat us to an iPhone?

The imminent arrival of such a product was a long-running rumour, though Apple denied it for just as long. Eventually Apple did team up with Motorola in 2005, but in the meantime other manufacturers were openly influenced by the way Apple designs, and others simply fantasized.

Marc Newson's talby phone for KDDI [p. 168] was frequently dubbed the answer to the Apple phone, as was (though less explicably) the Motorola V3 [p. 172]. But no one ever came so close as Isamu Sanada, a photographer who designs fantasy Apple products for a hobby. Not the only one – the Apple brand has inspired something of a cult that attracts this kind of behaviour – Sanada is definitely one of the best at it. Given that he created a design for a new laptop that predicted Apple's distinctive titanium powerbook months before it came out, perhaps a nice, shiny white music-playing HiPod is not so very far away.

Vodafone Toshiba V602T

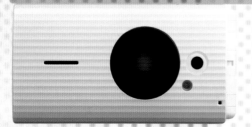

Another beautiful phone for Japan. Having been through the envy and frustration of being denied such products before, the rest of the world simply sighed collectively when informed of its release. The catchily named V602T from Vodafone KK was designed by Toshiba and it employed a whole new design language not seen before in phone design: texture.

Rather than different colours, the usual me of offering 'personal expression' with handse the V602T was available in metal and ceramic textural variations. Made of plastic, the effects were achieved using tiny dimples on the exter of the 'metal' phone, and a grooved finish tha vaguely reminiscent of porcelain.

The balance of straight and curved edges, and the circular motif for the external screen a

key layout, gave the V602T an interesting balance of stereotypically male and female design languages. With this handset, phones moved another step away from old-fashioned, male-driven electronics packaging to something altogether fresher and extremely appealing.

Pantech GI100

Korean manufacturer Pantech surprised everyone, especially the competition, by being the first to produce a biometric phone. Although it had consistently produced high-quality handsets, it was very much the new kid on the block.

Fingerprint recognition is about as futuristic as you can get after video phones (the ones where you see the person you are talking to, as opposed to video cameras), but no one expected this über-cool security device to turn up quite so soon. Previously the technology was only available in sci-fi films and on hand-held computers.

Aside from using a fingerprint scanner to lock and unlock the phone for authorized use, the GI100 also had an unusual feature called secret finger dial. Useful if you are a spy, a criminal or simply adulterous, this allocated up to ten secret speed dial numbers to your fingerprints. Users could tag it to a phone number that is not recorded in the phonebook and does not appear in the dialled numbers section after making a call.

The all-powerful biometric sensor sat in the middle of the four directional keys, in place of the more traditional 'OK' button. Not just a clever gadget, Pantech also tried to ensure that the biometric phone boasted a stylish visage. The grey-coloured clamshell that resulted was unfortunately fairly uninspiring. But if you're a spy, a criminal or having an illicit affair, that's probably quite a good thing.

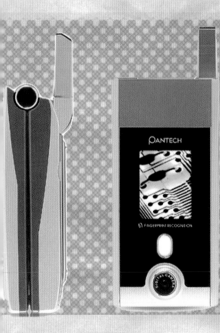

Samsung SCH-S310

The SCH-S310 is the world's first phone capable of three-dimensional movement recognition. Because it wasn't good enough that your phone could tell where you are, this one could detect how you are standing. While traditionally people control their phones by pressing keys or speaking slowly and clearly at it, Samsung's new technology was much more persuasive. To get this device to respond, all you had to do was gesture.

Samsung believed that three-dimensional movement-recognition technology would become an important new user interface. They developed the technology in a joint project between the company's electronics division and the Samsung Advanced Institute of Technology. It resulted in applications for twenty-two patents, and fourteen technical papers were presented at leading academic societies.

Built into the S310 was Samsung's new 'accelerometer'. This sophisticated technology uses sensors to recognize continuous movement in three-dimensional space, accurately calculating what is happening, then carrying out commands accordingly. The accelerometer knows what direction the phone is moving in and how fast, extending beyond the mere identification of a general position to include precise recognition of small and rapid movements as well as recognizing alphanumeric symbols.

The user writes a number in the

air in front of it. The phone reads this movement, recognizes it and then dials the number. Shaking the phone twice concludes a call or deletes spam messages. If you draw an 'O' or 'X' in the air the phone responds with a voice message 'yes' or 'no'. Move the phone sharply to the right and the selection on the integral MP3 player will move to the next tune on the list; move it to the left and the selection goes back one number.

Does this mean you could accidentally call your boss when you are out dancing? Theoretically, if you were doing the time-warp with your phone in your hand, you might cause some damage, but then you'd deserve everything you got.

The keypad's demise was already widely predicted, but three-dimensional movement recognition has the potential to change more than how we input phone numbers. It could transform the way games are played on a phone. Many functions can be controlled by movement instead of buttons, meaning that watching someone play games on the bus just got a whole lot more fun.

Predicting the future is a dangerous business, not least because people have a tendency to get it wrong. Film producers, cartoonists and authors are usually better at it, but perhaps that is because they don't usually predict so much as project.

This is especially true of phones – see the influence that *The Matrix*, *Star Trek*, Philip K. Dick and even *The Jetsons* have had. What began as fantasy is largely the reason we now have flip phones, video phones, biometric fingerprint scanning and other devices we never knew we needed.

But if we look to these sources to determine wh we can look forward to, what do we find? Not much seems the future, as we imagined it, is already her And all that's left is just around the corner: phones control your home with, unlock your doors with, CC your foyer via and run your bath for you are all in th pipeline. Watching feature films and video-casts or hand-held is not exciting any more, while the conc of a music player that holds all the songs in the wo is also no longer news, and having that capacity on phones will follow shortly.

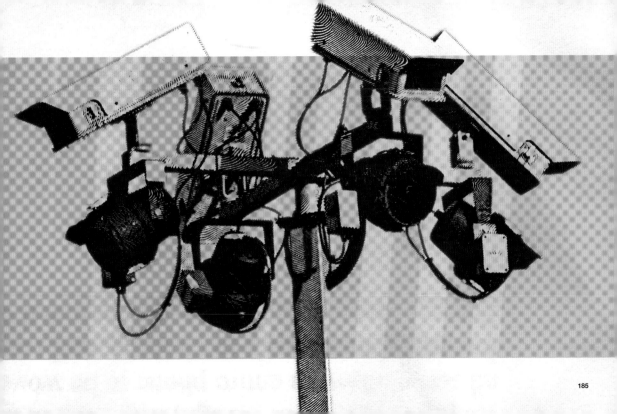

Possible Futures

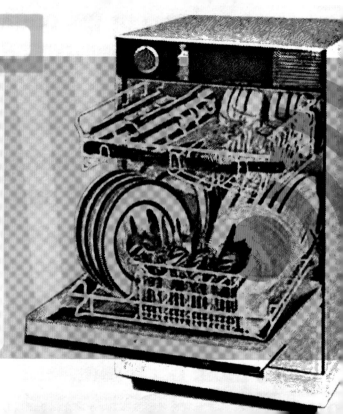

So, what of the more extreme ideas out there? When this book is languishing in a paper museum somewhere on-line in 2020 how might we be communicating over long distances?

The whole idea of a 'mobile phone' is, of course, already passé. They are called 'mobile devices'. This is the law of convergence. A phone is now a games console, a high-end digital camera, a television, a pacemaker, a dishwasher. There have been mutterings about a future in which we all carry a single mobile device that does everything. Depending on the source of the rumour, sometimes this device is government issued. But there are just as many people who maintain that convergence only serves a useful purpose up to a point before every function is compromised. The beauty of the iPod or Palm PDA is that they do one thing extremely well. Try to stuff everything into one, slimline, ergonomic and easy-to-use gadget and something is going to be

compromised, and you wouldn't want that to be your pacemaker.

Anyway, by 2020 we won't have any gadgets at all. By then our phones will be printed directly onto wrists or embedded in our brain, our teeth or other parts of the body. Before we get to that point, though, they will be embedded in our clothes. Intelligent textiles may allow an entire phone to be woven into a jacket without looking too ridiculous. Screens, meanwhile, can be incorporated into spectacles or even contact lenses as interfaces are developing to become ever more intuitive and less intrusive to our environments as well.

These wearable devices have to be clever as well as sartorially elegant for us to adopt them. The possibilities for our favourite future scenario, convergence, are more interesting with the onset of electronic fibres. Your T-shirt, for example, might monitor your health. If it senses that you are suffering a heart attack, the phone in your sleeve will automatically dial for an ambulance – giving your exact location details via satellite navigation. There are more marketable applications too: if you are stressed it will pump out soothing noises, or it might order you a pizza if you are hungry. An athlete's coach could monitor his charge's physical

condition during a race, just like F1 engineers can do with their cars.

This is just one step on from what is an imminent development in technology, when our phones become 'context aware'. Very soon we may have phones that get to know their user – you – and your daily routine. That way, claim the scientists, the phone will be able to offer all sorts of useful advice: instead of just sounding an alarm to remind you of an appointment, this next generation of phones will be able to make sure you don't drink too much the night before.

A context-aware phone could learn that whenever you phone at least three particular business associates, you will leave the office within an hour and meet them in a given place. It could presumably sort out an excuse for you as well, if it found out you had ignored its advice and stayed out drinking the night before. Worried about the government knowing too much about you? This is a phone that knows where you live.

Presumably they will work out a way to market it. Perhaps it will be through the useful application they are currently calling 'reality mining', when the phone can answer useful questions you might have about yourself, for example how long you spent working or partying last week, or when you last saw a friend.

Hopefully developments will continue to be dictated by what the users actually want, as they have largely done so far. Everyone in the business now knows that there is nothing more useless than technology without a user. Really practical solutions ought to be found in the next few years to the ecological problems of phones, making them easier to recycle and reuse, or making phones that we want to keep for longer, which can easily be updated with new functionality. Battery technology has a long way to go, as does the design of phone chargers (splashpads for all, please).

Ultimately, however, perhaps we shouldn't worry so much. According to one school of thought, in

Possible Futures

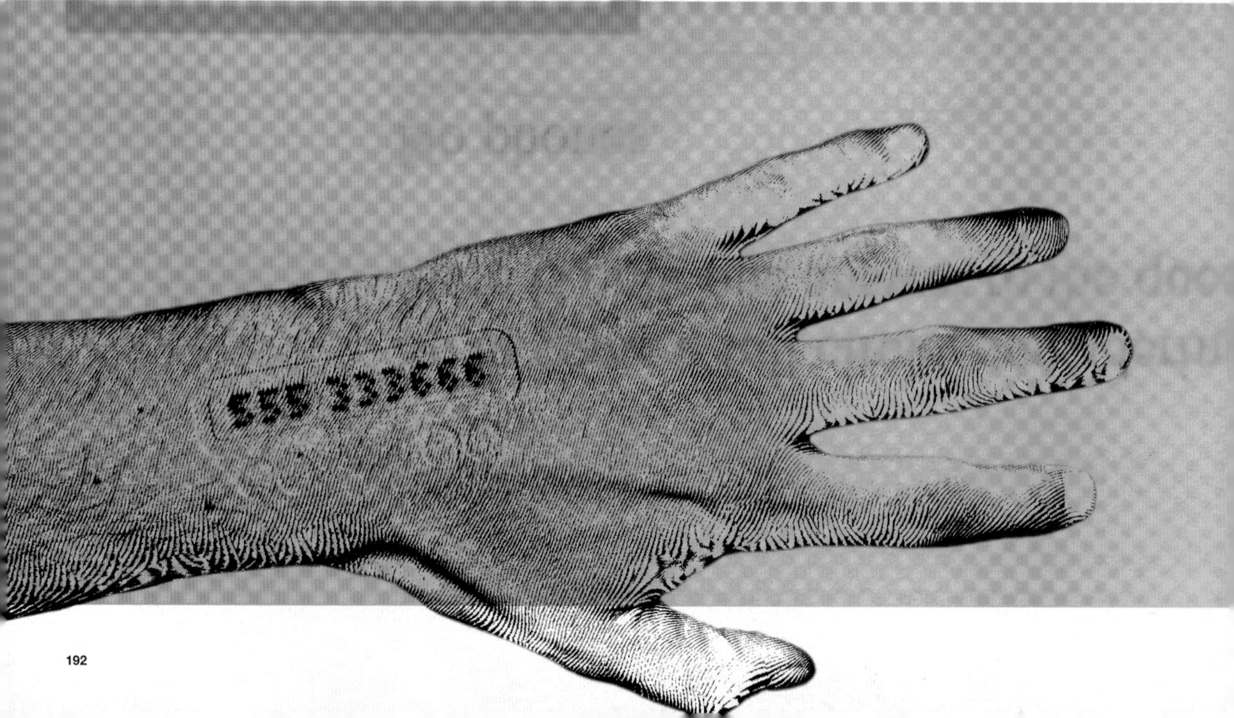

the future we will all be so connected, and phones so ubiquitous, that it will be the ultimate luxury to be uncontactable. This, apparently, is how we will be able to tell the rich and important people from the poor and lowly: the people who can afford not to have phones are the ones who really matter.

So, to come a full circle, considering wearables, embedded technology and convergence, there is only one answer to the question. What is the future for phones?

No phones.

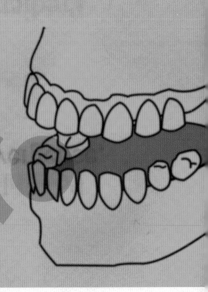

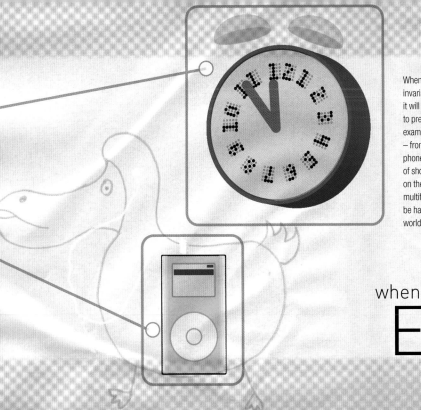

Whenever a new technology emerges, there's invariably a rush to speculate about what changes it will bring. A favourite pastime of journalists is to predict the end of life as we know it. Email, for example, allegedly spelled the end for many things – from paper in offices to the postal service. When phones became mobile, similarly, it was the end of shovelling handfuls of coins into phone boxes on the street. And, as phones become ever more multifunctional, there is plenty more imaginative fun to be had postulating what the next dodos of the gadget world will be.

when gadgets become...
Extinct

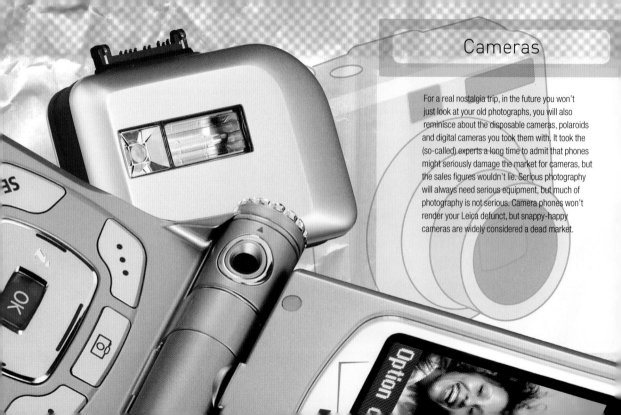

Cameras

For a real nostalgia trip, in the future you won't just look at your old photographs, you will also reminisce about the disposable cameras, polaroids and digital cameras you took them with. It took the (so-called) experts a long time to admit that phones might seriously damage the market for cameras, but the sales figures wouldn't lie. Serious photography will always need serious equipment, but much of photography is not serious. Camera phones won't render your Leica defunct, but snappy-happy cameras are widely considered a dead market.

Keys

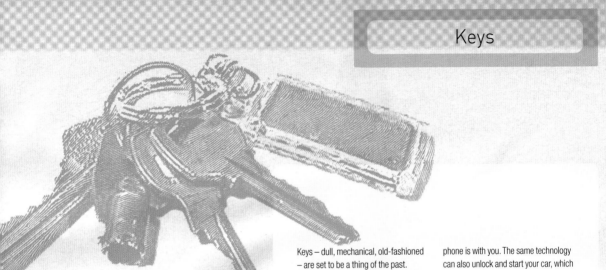

Keys – dull, mechanical, old-fashioned – are set to be a thing of the past. Instead, tap your pin into your phone as you approach your front door and it will swing open. In the not-too-distant future, you'll be able to open your front door from wherever you are, as long as your phone is with you. The same technology can also unlock and start your car, which may be available from 2007, pending the agreement of car manufacturers who may be reluctant to relinquish control over security.

Remote Controls

Good for more than just channel hopping, use your new remote control for calling friends! A phone is far superior as a remote than the remote ever was: it is the ultimate couch-potato accessory. Phones can be programmed to control just about anything: lights, curtains, alarms, air conditioning. It will tell the stereo to play your favourite song as you walk in the room and tell your computer to check your emails when you are near (unless it's already done so). Lean back on the sofa and relax. Laziness is no longer an option, it's a way of life.

Beaming business cards between phones and PDAs was once something that only happened between the geekiest people at the geekiest trade fairs. But these people know a good technology application when they see it, and their enthusiasm resulted in more phones with better networking capabilities. Phones now scan business cards as well as send them, all of which encourages people to store their contacts in this way, leaving the little black book and business-card holder for traditionalists and technophobes only.

Another paper-based eccentricity, the postcard, is dwindling too, replaced by the DIY electronic postcards that are photo and video messages.

Clocks and Watches

Phone handsets have always told the time. They also wake you up in the morning and time your eggs cooking, leading many people to conclude that burdening their wrists with superfluous gadgetry is pointless. But watches have never been just about telling the time, they are also status symbols and fashion accessories. They have also evolved into remote controls, maps, compasses, pagers and personal assistants – all things that phones can do too. Watch makers and phone makers have been engaged in constant battle since the latter's first appearance – watch designers want theirs to be the device of choice, just as phone people do. It seems that phones are winning. They have the edge for being new, for being bigger, for being phones.

Landline Phones

Although most people aren't actually swapping landlines for mobile phones, preferring to juggle the benefits of each (mobiles are flexible, landlines are cheap), eventually traditional landlines will be obsolete. The arrival of voice over IP (VoIP) connections, which allow an internet connection to route phone calls across a network extremely cheaply, will speed up the landline's demise.

Other ancient telephony products that are on the way out include answer phone machines and phone booths, though what's to be done with them remains a problem. Given the advancing capabilities of phones, the internet-booth idea will soon be defunct.

Once upon a time people used to write telephone numbers and appointments in a note book, with actual paper pages, that was intended specifically for this purpose. For many it was a little black book, for others it was adorned with pictures of kittens. Some people kept all they needed to organize their lives in a ring-bound device known as a FiloFax, while others went so far as to memorize telephone numbers. But the arrival of personal digital assistants and phones with calendars, speed-dial keys, and in-built contacts books meant there was no longer any need for such shenanigans. Slowly these products disappeared from handbags and pockets everywhere. Perpetual contact, as provided by mobile phones, means that appointments themselves are also steadily becoming a thing of the past.

Cash

© BCE ECB EZB EKT EKP 2002

It is likely that the handbags of the future will be very light indeed, as it's not just the notebooks that can be thrown out, but wallets too. Embedded with smart chips such as Sony's FeliCa, phones can increasingly be used in place of membership cards, ID cards, electronic plane, train or parking tickets and, of course, cash and credit. When Japanese manufacturer NTT DoCoMo unveiled the first line of phones with this capability in 2004, it took the first step in its plan to become a 'life infrastructure company', which says it all.

20 EURO

Portable Games Consoles

Specialized hand-held machines, such as games consoles, PDAs, pagers and GPS devices, all face extinction in the face of the phone as all-powerful entertainment-communication-information device. Of these, the games console has been most prolific. Although GameBoy and others have incorporated some phone functions, such as text messaging, into their own products, they are fighting a losing battle. Phones, with far superior functionality, portability and accessibility, will trump consoles almost every time.

While the iPod was the biggest success in consumer gadgetry since the Rubik's Cube, and the biggest innovation in music since the Beatles, it would have to do a whole lot more to survive the rampant march of all-doing phones. A phone is now also a music store, and this will affect iPods as much as iPods affected Walkmen and personal CD and minidisk players. The potential for a mobile device as a music delivery platform is obvious, given that there are hundreds-of-millions of phone owners in the world – a considerably larger potential market than the millions of people who use iPods or other MP3 players. The wallet function of future phones is also handy for impulse buying tunes at any time.

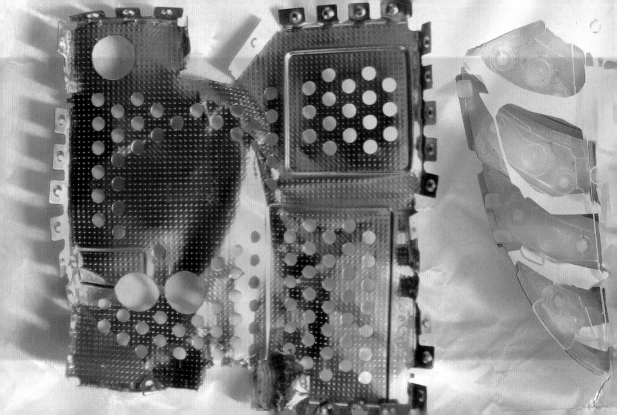

The Environmental Impact

Even though most phones are built to last six years or more, manufacturers and service providers can be extremely persuasive in getting people to change their phones every six months. With technology advancing so rapidly, phones can seem obsolete even before they've stopped ringing.

Given that some phones and their accessories contain substances that are amongst the ten most dangerous known (lead, for example, and arsenic), it's important to consider what happens to phones we don't want any more. The more valuable materials commonly used in phones – gold, silver, etc. – can also prove a pressing argument.

Most redundant phones, however, end up adding to the huge swelling volume of toxic electronics in land fill. In California, one environmental group estimated that since the start of the new millennium residents have been throwing away an average of 45,000 phones every day. Nationwide, it is estimated that about 130-million

handsets end up in US land fills or incinerators annually – that's about 65,000 tons of waste.

America is easily the worst offender in most matters of waste and, despite many well-meaning schemes on a local level, lack of interest from federal government has meant that less than one per cent of used phones are recycled or reused. Thankfully, this is not true of the rest of the world: the figure is way behind Europe and Japan, for example, where land-fill space is increasingly precious and manufacturers have discovered profitability in recycling.

A growing number of companies offer refurbished phones to their prepaid wireless customers, or re-sell them in developing countries. But if a phone is broken and cannot be repaired or if – like most – it has simply been superseded by a more exciting, younger model, recycling is the obvious answer.

Breakdown of a phone:

- Printed circuit boards are made from mined raw materials, including beryllium, coltan, copper, gold, lead, nickel, tantalum, zinc and other metals. These are generally ground, heated, treated, shipped and finally manufactured and mounted on the board itself.

- The board is constructed from plastic, which requires crude oil and chemicals to make, and fibre glass (requiring sand and limestone), and is then coated with a thin layer of gold. The electronic components and wires are soldered to the board and secured with protective glues and coatings.

- The liquid crystal display (LCD) contains various liquid crystalline substances, such as mercury, sandwiched between glass or plastic.

- Phones can use several different types of batteries: nickel-metal hydride (Ni-MH) or nickel-cadmium (Ni-Cd), which contain nickel, cobalt, zinc, cadmium and copper; lithium-ion (Li-Ion), which uses lithium metallic oxide and carbon-based materials; or lead acid.

- The electricity used in charging the world's phones is not insignificant either, though next to cars and houses they are by no means the biggest drain on resources. Even so, Motorola, in collaboration with Freeplay, have developed FreeCharge – a gadget for recharging batteries using muscle power. A hand-powered device, it can provide fifteen minutes of talk time after just three minutes of squeezing a generator. We're also still waiting for hydrogen fuel cells and zinc-air and solar-powered batteries to conserve natural resources and reduce waste.

The Environmental Impact

The Environmental Impact

en a phone is recycled, typically it will be shredded and
tals and plastics are separated. The metal shreds are
arated into different fractions. Aluminium, ferrous metal
copper can be sold to metal refineries, where they are
elted and purified for reuse. Instead of poisoning the fishes
he sea, metal from handsets that are recycled might be
orporated into a new frying pan, for example.

ted circuit boards are smelted and treated using
trolysis to separate copper from other materials.
ycled copper is often used for plumbing pipes, while the
aining metallic sludge containing gold, palladium and
inum will go to a precious metals plants for recovery.

stics, the principal material in the majority of phones,
't be recycled so easily. But they can be burnt and, along
other 'difficult' materials, they are sometimes used as
in the metal recovery processes. If they are shredded,
ever, the ugly plastic can still be used in a diverse range
ems, such as plastic benches, fences, car wheel trims,
ter cartridges or traffic cones.

- About sixty-five to eighty per cent of a phone can be recycled
  and reused. That figure is even higher if the energy is
  recovered by using the plastics as fuel.

Part 3 **Technology Focus**

What's Inside?
Case Study: Samsung Hera
Case Study: KDDI au Project
Case Study: Vertu
Case Study: Nokia Fashion
Case Study: Motorola V3 RAZR
Case Study: Sony Ericsson P-Series
Text Messaging
Connectivity
Ringtones
Camera Phones
The Rise of 3G

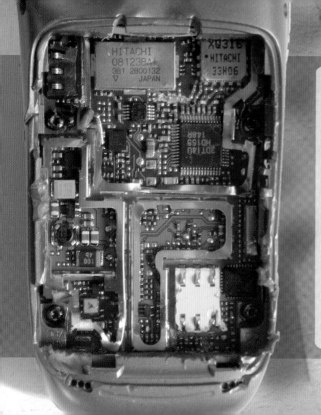

What's Inside?

The little green men who live inside your television set won't fit into even the biggest phone, so engineers had to find other ways to make them work. Your handset contains:

microphone Captures the caller's voice.

speaker Through which someone else's voice is heard.

LCD display Shows useful information, such as who is calling and what the time is.

keypad The traditional way to control a phone, although with touch screens and scrollwheels, it is no longer necessary.

battery Powers the phone.

digital signal processor Processes▮ speech and transmits and receives sig▮

CODEC The chip that compresses and decompresses analogue sound into dig▮ transmit it. It does this very fast.

Radio Frequency unit Changes voi▮ or data into radio waves, which is sent ▮ the nearest cellular base station. Incom▮ information comes back via the same r▮

antenna Although pull-out steel rods▮ no more, and antenna systems are usu▮ internal, every phone needs an antenn▮ transmit and receive radio waves.

SIM card reader Reads subscriber information on the SIM card and transr▮ digitally to a network via the RF unit.

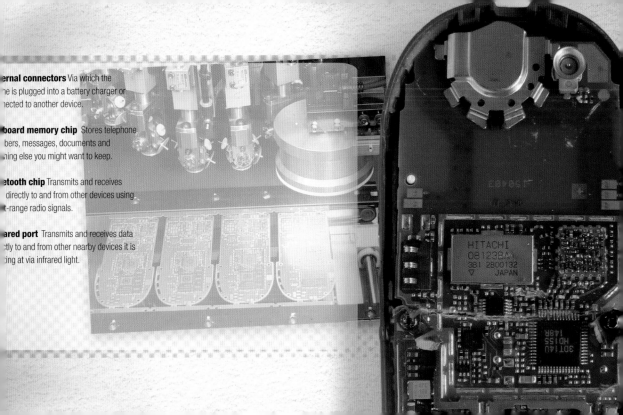

ernal connectors Via which the
ne is plugged into a battery charger or
nected to another device.

board memory chip Stores telephone
bers, messages, documents and
hing else you might want to keep.

etooth chip Transmits and receives
directly to and from other devices using
t-range radio signals.

ared port Transmits and receives data
tly to and from other nearby devices it is
ting at via infrared light.

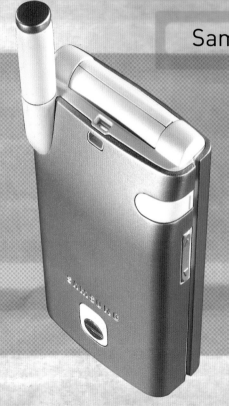

Samsung Hera

The Hera phone is a neat, simple compact phone designed by Samsung Design Europe for Samsung in 2003. Intended for the European market, the Hera was initially released in Korea.

Samsung has been a design-led company since 1996 when its chairman Kun-Hee Lee announced the year of the design revolution. One of the company's main objectives has been to use design to create products that enhance people's lives. Phones, which have always had potential in this area, have become one of Samsung's core businesses.

The Korean-based company (whose name translates as 'three stars') made its first phone, the SH-700, in 1993 and continued to innovate, securing its position as number two, behind Nokia, in the Asian market by 2000. Samsung attributed its success to achieving a balance between the sophisticated

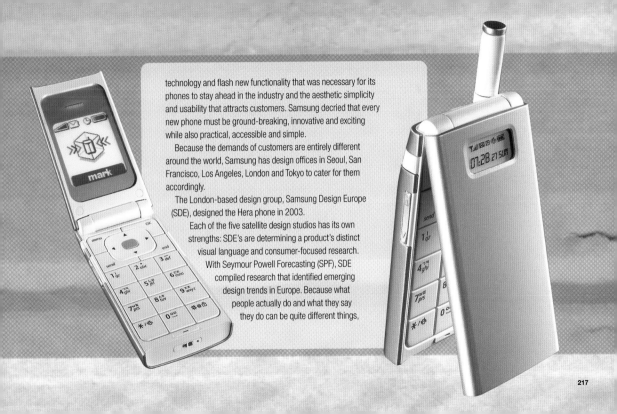

technology and flash new functionality that was necessary for its phones to stay ahead in the industry and the aesthetic simplicity and usability that attracts customers. Samsung decried that every new phone must be ground-breaking, innovative and exciting while also practical, accessible and simple.

Because the demands of customers are entirely different around the world, Samsung has design offices in Seoul, San Francisco, Los Angeles, London and Tokyo to cater for them accordingly.

The London-based design group, Samsung Design Europe (SDE), designed the Hera phone in 2003.

Each of the five satellite design studios has its own strengths: SDE's are determining a product's distinct visual language and consumer-focused research. With Seymour Powell Forecasting (SPF), SDE compiled research that identified emerging design trends in Europe. Because what people actually do and what they say they do can be quite different things,

SDE and SPF spent time with potential customers in their homes and work places and watched how they used the products they came into contact with.

One outcome was the realization that too many phones couldn't be used by many of the people they were supposed to be designed for. They were too small; the buttons were too fiddly; the interfaces were wrong; and they were increasingly packed with functions that consumers didn't want, didn't need and wouldn't ever use but had to pay for.

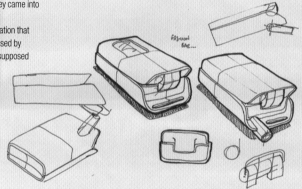

It seemed that manufacturers were so busy fi problems for all the new technology to solve, th had forgotten that the solution was sometimes leave it out altogether.

SDE and SPF identified a new consumer ⬤ which they called 'emotional minimalism'. S objective was to design a European emoti minimal phone, essentially a phone devoi of unnecessary technology. While the competition's products were becoming m and more visually complicated, Samsung identified a user who just wanted a phone they just made calls with. A phone that was quite simply, a phone.

SDE wrote a brief to this effect, mocked-u some initial concepts, produced models and presented them to Samsung. The board

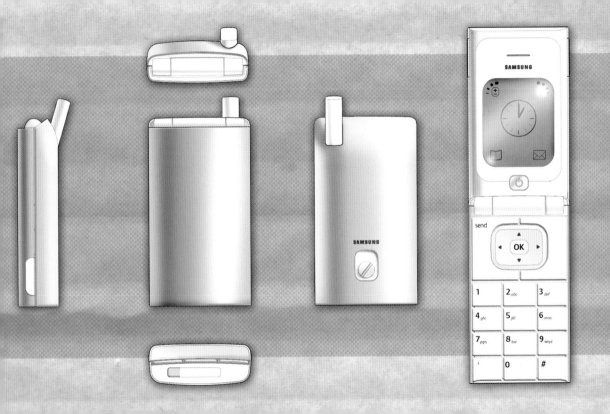

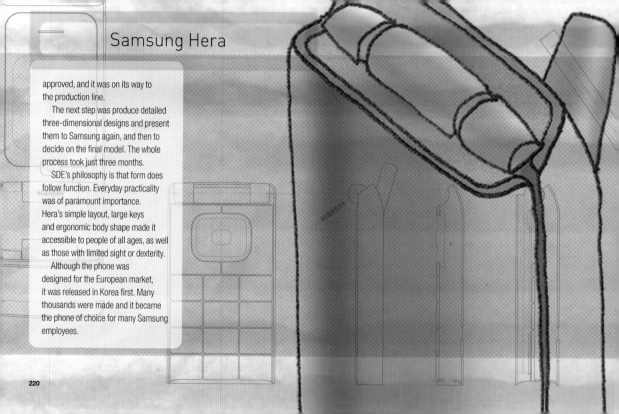

Samsung Hera

approved, and it was on its way to
the production line.

The next step was produce detailed
three-dimensional designs and present
them to Samsung again, and then to
decide on the final model. The whole
process took just three months.

SDE's philosophy is that form does
follow function. Everyday practicality
was of paramount importance.
Hera's simple layout, large keys
and ergonomic body shape made it
accessible to people of all ages, as well
as those with limited sight or dexterity.

Although the phone was
designed for the European market,
it was released in Korea first. Many
thousands were made and it became
the phone of choice for many Samsung
employees.

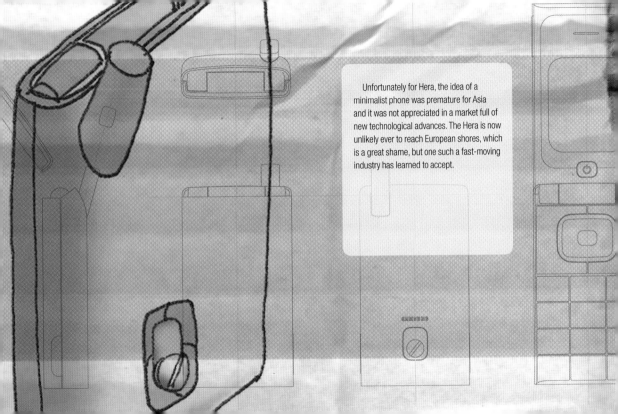

Unfortunately for Hera, the idea of a minimalist phone was premature for Asia and it was not appreciated in a market full of new technological advances. The Hera is now unlikely ever to reach European shores, which is a great shame, but one such a fast-moving industry has learned to accept.

SAMSUNG

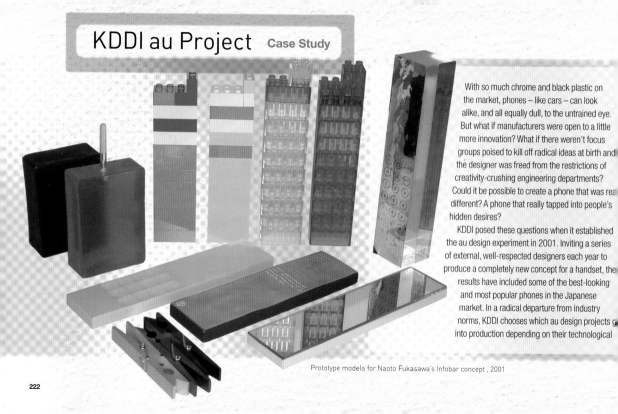

KDDI au Project Case Study

With so much chrome and black plastic on the market, phones – like cars – can look alike, and all equally dull, to the untrained eye. But what if manufacturers were open to a little more innovation? What if there weren't focus groups poised to kill off radical ideas at birth and the designer was freed from the restrictions of creativity-crushing engineering departments? Could it be possible to create a phone that was really different? A phone that really tapped into people's hidden desires?

KDDI posed these questions when it established the au design experiment in 2001. Inviting a series of external, well-respected designers each year to produce a completely new concept for a handset, the results have included some of the best-looking and most popular phones in the Japanese market. In a radical departure from industry norms, KDDI chooses which au design projects go into production depending on their technological

Prototype models for Naoto Fukasawa's Infobar concept , 2001

feasibility and the market's reaction to the concept designs once they are finished.

So what do people really want from their handsets? Judging by some of the phones that KDDI has put into production, a resemblance to a peeled potato goes down especially well, as does chocolate bar packaging.

The first designers to take part were Naoto Fukasawa and Takahashi Nikaido. Besides setting up the IDEO design company in Japan, Fukasawa sat on the design board at Muji and was responsible for the company's famous wall-hung CD player in 1999. Among many achievements, Nikaido helped to develop Casio's G-shock watches.

KDDI au Project

The two designers produced a number of concepts, though Fukasawa's Infobar concept [p. 150] was the only design from this year to go into production. It was sold in a wrapper reminiscent of a bar of chocolate. Nikaido's concepts were more conventionally prophetic. The Rotary, for example, was a phone with a screen that could be positioned at any angle. With his wearable phone, he investigated new portability options with bracelet- and pendant-style options.

Fukasawa, enthused by the Infobar's success, designed another concept for the au design project the following year: the Ishicoro (Japanese for stone). Fukasawa noticed how people played with their phones even – or especially – when they weren't using them, and he likened this action to how people subconsciously behave if they have a shiny stone to hold. Believing that the irregular surfaces of a real stone, as opposed to an artificial alternative, would better inspire an emotional connection with nature, Fukasawa went to a river and found an appropriate rock to take a cast from.

The stone phone glows with an incoming signal and is a wonderfully intuitive, emotionally aware design. Sadly, Ishicoro was never put into production. It is, however, remarkably similar to

Ichiro Iwasaki's concept for Grappa 00.

224

ukasawa's Ishicoro, or stone phone, prototype, 2002

another Fukasawa design that did: the W1
otherwise known as the 'peeled-potato'
phone. According to KDDI, the W11K is no[t]
evolution of Ishicoro, but the idea is very m[uch]
the same. Rather than a polished stone, th[e]
faceted form of the W11K seems a little ea[sier]
to manufacture, while still being a touchy-
feely product. KDDI maintains that Fukasa[wa]
drew inspiration from the visual and tactile
qualities of washing a freshly peeled potat[o]
in water.

Other 2002 projects included the Grap[pa]
and Grappa 002 phones by Ichiro Iwasaki,
a renowned Japanese electronics designe[r.]
Grappa was an extremely sleek phone with[]
a leather case and Grappa 002 had a slide[-]
open handset.

Another 2002 design, the Apollo by
progressive multimedia artist Ichiro
Higashiizumi, was a concept that tried

Ichiro Higashiizumi's Apollo 02, 2002

to be all things to all people. Compact but scary-looking, it featured a full keyboard that could be used like a videogame controller. Higashiizumi realized how spacey Apollo had looked and took it a step further with Apollo 02. This was how a phone would have looked if *2001: A Space Odyssey* had featured one. The LCD display even had interchangeable tiles that you could use to write text messages on.

The third au design project model to go into production was Marc Newson's talby [p. 168].

Simple, clean, modern and just plain gorgeous, the talby 'has everything you want (and nothing more)'. The name came from a character in one of Newson's favourite science-fiction films, *Darkstar*. 'He is completely absurd but both cool and relaxed. More importantly though he is incredibly funny. These are qualities I always look for.' Judging by the sales, he's not the only one.

Higashiizumi's Apollo, 2002

Vertu

tu, the $10,000 phone, was conceived in
95, by Frank Nuovo, then head of design
Nokia. Inspired by the idea of a phone as a
tus symbol, he compared it with a luxury
tch or fountain pen – products that had
olved from functional tools to objects of
ecision and status. Nuovo made a formal
oposal to the Nokia board in 1997 with
ter Ashall, now senior vice-president
product development and strategy at
tu. The early concept and planning work
gan soon afterwards and the Vertu Signature
s launched in a limited edition of 1,000
2002.

There is a luxury end to most mature product
eas, says Nuovo. 'The Vertu concept simply
cognizes that a person wearing a fine watch,
iting with a fine writing instrument and
earing an Armani suit would want continuity of
ality with his mobile phone.'

While the value of luxury products is
often found in time-honoured traditions and
craftsmanship, phones had no such history.
Instead, says Nuovo, 'Our work focused on
applying the principle of obsessively crafted
to everything about Vertu. The result was the
Vertu Signature, the "grand complication"
of mobile phones'.

That's an understatement. The Signature
featured 416 mechanical parts; underwent over
100 rigorous tests; and its creation resulted in
74 new patents. Authentic materials, uniquely
crafted construction and finishing methods all
add to a luxurious design that was exclusive to
Vertu. Nuovo, who had been designing mass-
market phones for a decade, had explored 'all
the easy possibilities' of materials. Finally he
could use pure metals, sapphire, leather and
ceramic, all of which had been faked using
plastic approximations in phones in the past.

Vertu

'Authentic materials do not wear out – they age gracefully,' he says. The sapphire crystal face does not scratch, while highly polished ceramic, of the kind used on the exterior of space shuttles, could be moulded into a gently curved, warm and scratch-resistant ear pillow and battery cover.

The design and engineering teams at Vertu travelled extensively in search of suppliers and craftsmen, challenging prestigious luxury-goods manufacturers to apply their fine and fiddly ways of working to the high-tech world of communications technology. Both sides soon discovered that it was not an easy task. From the Swiss company supplying gold bezels to the French factory cultivating sapphires for its face, these manufacturers had to rewrite the rules set by smaller mechanical watches. 'When no sapphires could be found big enough to fit

the face of the Vertu Signature, the factory employed a new method to grow them larger – making Vertu's sapphire crystals one of the largest such gems ever made.'

The keypad alone contains over 150 different parts and each stainless steel key sits on two extremely hard-wearing, jewelled bearings to produce a precisely controlled, fluid movement, which has been tested to over two million presses. The key graphics are laser cut micro-perforations through which backlighting, which is tinted to match the colour of the leather case, shows. Over 575 holes are drilled into every keypad and each is injected with a special resin using a 0.3-mm syringe to prevent moisture and dirt from entering.

If the Signature was the luxury watch, the second Vertu model, the Ascent, was the luxury car. It was constructed from extremely

hard, precision-engineered metal alloy called Liquid, which was custom-develop for Vertu. 'It is harder than titanium and it un-regimented atomic structure helps giv durability and scratch resistance,' accord to Nuovo.

The excuse for such extravagance is that it offers durability and an extra-long life. But technology moves fast, and so does fashion. So, where possible, Vertu has designed the phones to enable updates. When the Vertu Signature was first launched, the display interface was grey-scale.

Vertu's Ascent model took its influences from the luxury automotive industry with hand-stitched leather panels, 2002

But when colour screens became available, owners could simply upgrade their handset. Plus, says Nuovo, the design is essentially a modular construction so its parts can be serviced or replaced easily.

The components are shipped to the Vertu factory in the UK where they are finished and hand-assembled. Parts are hand-polished, and hand-texturing techniques, such as brushing and bead-blasting, provide a range of surface finish details to the parts. Even the SIM card retainer has a special secondary finish. Assembly by hand then takes about four hours.

There are many who will be incredulous at such extravagance, but there are plenty of people who are more than willing to pay for such luxury. Nuovo, who spotted that first, is onto a winner.

Nokia Fashion Case Study

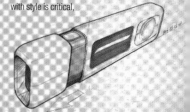

People, curious creatures that they are, like to express their individuality through their phones, as interchangeable covers, ringtones and wallpapers all demonstrate. Nokia has pioneered phones as style and status symbols ever since it launched the 8210 in 1999 [p. 132], and is a trailblazer too of the idea of phones as fashion accessories. The 7260, 7270 and 7280 formed a collection of catwalk-inspired, prestige phones that was launched in 2005.

The previous year the company had made a bold move by introducing exotic fabrics into phone design: the Nokia 7200 included a limited edition in white leather. Although industry pundits were sceptical about Nokia's directional but frivolous fashion agenda, the company pursued it determinedly.

The success of the three phones, particularly the 7280, and their accessories – including cases and changeable covers – was a turning point in phone-design history.

Fusing consumer electronics with fashion is always tricky, too often it is executed by product designers with no idea of the nature of trendsetting, who mistake candy colours, whimsical shapes and customizable fabric for haute couture. So what had Nokia learned that other manufacturers hadn't? 'Consumer understanding is key,' explained Tanja Finkbeiner, senior designer at Nokia. 'To be daring with style is critical,

Sketches for the 7280, 2005

however it is equally critical to ensure that the phones' functionality and usability is appropriate to consumer expectations. The on consumer's desire to stand out in the crowd, oticed and set new trends is a key purchasing r, but if the phone is not packed with the latest nology it can be easily dismissed as a gimmick.' ccessories are often the key to fashion. important thing with the phone as a fashion ssory is how you wear and carry it, you hold it and what you do it when you are

not using it,' says Finkbeiner. 'The style of the phone is reflected in all of these actions and, much like belts, handbags, shoes, jewelry, watches etc., the phone becomes part of our individual style.'

But wearability alone does not make a fashion phone. 'Our products are also considered aspirational,' says Alastair Curtis, design director of Nokia's mobile phone division. 'They often evoke a sense of wonder, making people go "how did they DO that?", as well as a sense of pride and joy in ownership.'

The iconic 7280, or the lipstick phone, was the most eye-catching of the three phones in Nokia's fashion collection and it was pivotal to

Nokia's continuing success. The brief was for it to be small, sexy, eye-catching and cool. 'It had to challenge consumers' expectations of what a phone can be and how it is used. The first question we wanted people to ask was "that's cool, what is it?" Once shown how it comes to life and how it works we then wanted them to say "I want one".'

Nokia Fashion

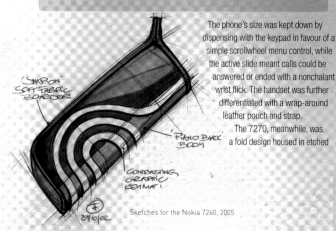

'WRAP ON'
SOFT FABRIC
BOARDER

'PIANO BLACK'
BODY

CONTRASTING
GRAPHIC
KEYMAT !

Sketches for the Nokia 7260, 2005

The phone's size was kept down by dispensing with the keypad in favour of a simple scrollwheel menu control, while the active slide meant calls could be answered or ended with a nonchalant wrist flick. The handset was further differentiated with a wrap-around leather pouch and strap.

The 7270, meanwhile, was a fold design housed in etched stainless steel with a customizable, detachable felt wrap. The 7260 was available in a white 'lacquered' finish with chrome and steel accents in an Art Deco motif. Like the others, it also came with its own fabric pouch.

Whereas traditionally making something fashionable meant making it appeal to the female market, these phones were designed to appeal to men and women.

This was crucial. 'The design is meant to dare the consumer to be different. It was important that we never compromised usability for the sake of style.'

The designers were also aware that there was a particular audience for fashion phones. This collection was targeted at an audience that was inspired by the latest fashion trends. 'The Art Deco–inspired geometric designs and colours, which was a strong theme in the fashion world at the time, would appeal directly to the fashion-conscious crowd,' says Finkbeiner. And the mirror on the front? That's especially considerate of this audience too.

The Nokia 7270 was the fold design in the trio of Art Deco–inspired phones, 2005

Motorota V3 RAZR Case Study

In July 2003, two Motorola teams – one of engineers and the other of consumer-experience designers – met at the company's world headquarters in Libertyville, Illinois. Their mission was to create the thinnest clamshell phone Motorola had ever developed.

The phone must be no thicker than 14 mm (approx. ½ in) when closed, a challenging target requiring a revolutionary design and a leap forward in phone technology – if it could be achieved at all.

A year later in Copenhagen, Motorola unveiled the V3, a stunning aluminium clamshell phone that was just 13.9 mm thin.

Although the silver good looks implied a design-led approach, it was a major breakthrough from the engineering team that guaranteed a crucial space-saving internal configuration for the phone's circuit board and allowed it to happen. They decided that, rather than stacking the internal components, they could be laid out along the same plane, making an ultra-thin phone a possibility.

They still needed completely new antenna, hinge and keypad designs. The materials were crucial. According to chief engineer Roger Jellicoe (the designer was Chris Arnholt), magnesium was a key choice for housing the internal components as it could be moulded into very thin sections and was eighteen-times stiffer than traditional plastics.

Other key materials were aircraft-grade aluminium, used on the external housing, and a nickel finish to the surfaces of the open clamshell for a subtle tone-on-tone shading effect. Less than a millimetre of height was available for the lenses for the external display and camera so the team sourced chemically hardened glass; plastic would not have been strong enough.

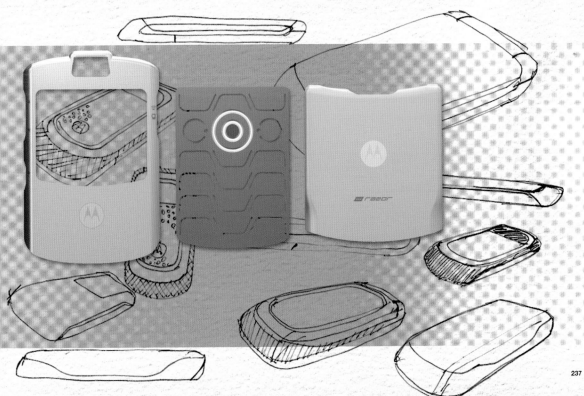

Motorola V3 RAZR

Over an intensive four-month period, the V3 went from crazy idea to tangible concept. The design team began to sketch the form of the Motorola V3, including the potential shapes that a razor-thin keypad could take. These drawings were turned into three-dimensional models, which were carved and sculpted to refine the shape and styling.

Aluminium models were made at key stages in the design and, as the V3 began to emerge, the designers and engineers used CAD software to produce a digital model and to mould the shape around the internal components, define the design and reduce the volume even further. The design of a central hinge was another breakthrough. Nicknamed the 'knuckle', it was designed to keep a consistent visual line between the keypad and the display sections.

Meanwhile, five engineers were each given a week to come up with two concepts for an internal antenna design. The winning concept not only stayed within the key dimension, but its performance has proved to be one of the best in Motorola's portfolio.

The keypad was one of the last elements of the Motorola V3 to be finalized, partly because it presented one of the biggest problems. The space available for the keypad measured just a third of the standard allocation. The solution, the slimmest keypad the company had ever made, was a single sheet of nickel-plated copper alloy with the numbers and characters chemically etched into its surface. Below it, a blue electro-luminescent panel defines each key, illuminating the figures.

Once the phone had been designed and prototypes were made, testing began.

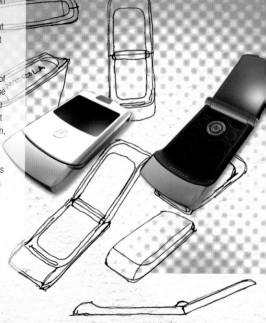

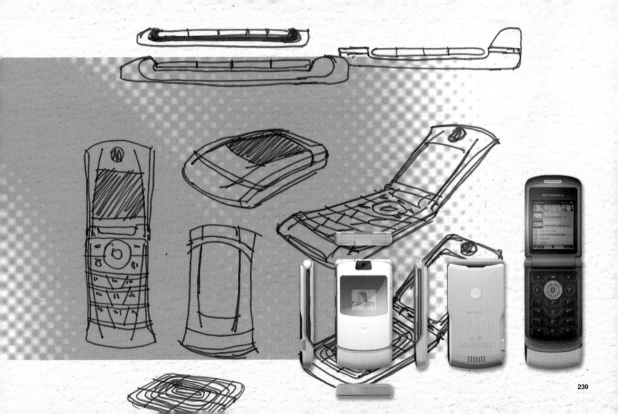

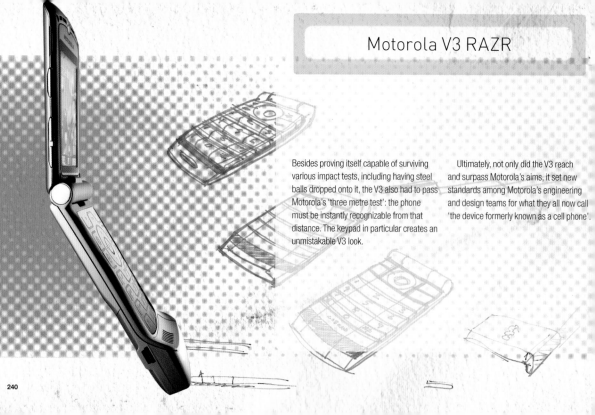

Motorola V3 RAZR

Besides proving itself capable of surviving various impact tests, including having steel balls dropped onto it, the V3 also had to pass Motorola's 'three metre test': the phone must be instantly recognizable from that distance. The keypad in particular creates an unmistakable V3 look.

Ultimately, not only did the V3 reach and surpass Motorola's aims, it set new standards among Motorola's engineering and design teams for what they all now call 'the device formerly known as a cell phone'.

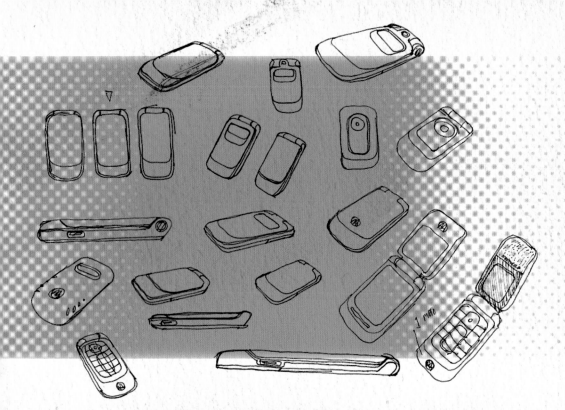

1 mm

Sony Ericsson P-Series

Sony Ericsson's P-series of smartphones has evolved over several years, since Sony and Ericsson joined forces in 2001. Both companies urgently needed to increase handset sales, and they knew they'd have to come up with something very sophisticated if they were going to compete with Nokia and Motorola.

The original Sony Ericsson smartphone, the P800, was created in the Sony Ericsson Creative Design Centre. It quickly evolved into the P900, and subsequently the P910, the P910i, and so on.

The Creative Design Centre is a multidisciplinary hotbed of creative professionals, including industrial designers, colour and material designers, human interface designers and graphic and packaging designers. It has design studios in Sweden, the USA and Japan, and everything from the operating system to the packaging is created in this environment.

According to Henrik Jensfelt, lead designer for smartphones at Sony Ericsson, when they designed the P800 they wanted to set the standards for the – very new – smartphone category, and create a flagship product for the brand.

A brief was written to design a smartphone capable of all sorts of ingenious wizardry and fun. A few rough sketches were done to determine the 'master idea' and then the team began to create numerous three-dimensional variations of the product digitally. At this stage the colour and material designers add their magic before the human interface designers deliver the graphical appearance of themes, icons and wallpaper. The packaging and graphic designers are let loose last of all.

Sony Ericsson P800, 2003

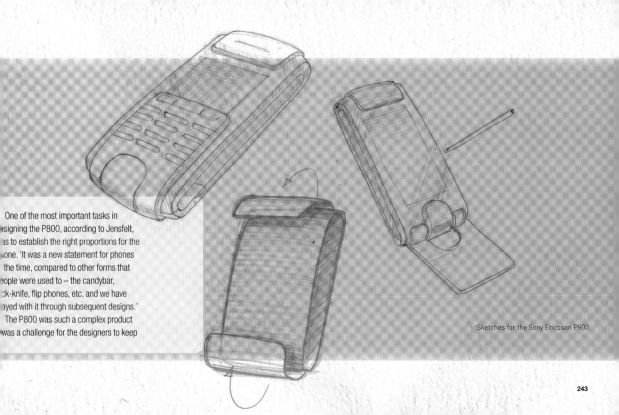

One of the most important tasks in
designing the P800, according to Jensfelt,
was to establish the right proportions for the
phone. 'It was a new statement for phones
at the time, compared to other forms that
people were used to – the candybar,
lock-knife, flip phones, etc, and we have
played with it through subsequent designs.'

The P800 was such a complex product
it was a challenge for the designers to keep

Sketches for the Sony Ericsson P900

Sony Ericsson P-Series

the form as simple and usable as possible. 'The palm size was very important to keep because as soon as the product gets too big it's not going to work anymore as a mobile. You have to think about the way the product will be used.'

The new smartphone was very well received. But the speed of technology meant that the P900 evolved soon after its release, with dramatically better screen capabilities and a slightly slimmer, lighter form. The light blue plastic casing had evolved into a darker, chic metallic blue. A year later, and the P900 upgrade, the P910 emerged. Offering far improved input possibilities, including a tiny QWERTY keyboard, the P910 was available in a sleek, silver metallic finish.

According to senior design manager Michael Henriksson, who led the colour, material design and communication team, this was because they realized that smartphones were as popular with women as they were with men, and it was necessary for the design to appeal to both. 'In the P800 the pale blue colouring reflected the new statement that we were making with this flagship phone. In the P900 the darker, moderate blue, was – like the phone itself – more sophisticated. By the time of the P910 we were using metallics and a more exclusive finishing that would appeal to everyone.'

The challenge with the subsequent iterations of the smartphone range has been to incorporate all the necessary new functions and capabilities while keeping

the look of the phone approachable and consistent. As Henriksson explains, it is about sending out a very strong and simple message, and not letting new technology compromise that. 'People use their phones in different ways,' said Jensfelt. 'You have to put yourself in the user's shoes – and think about how the product can reflect the ways it is used best.'

For example, the huge screen on the P900 and P910 was a crucial feature to emphasize. 'You look at what is important and then make those the key elements of the product design. It's very much about how you see the product emotionally.'

Sony Ericsson has a dedicated team for usability testing, which is an expensive process when you have

as much functionality as a smartphone; the P800 testing cycle lasted four months. 'But you are always building on what you've already done,' explained Jensfelt. 'With the P900 we had done much of the testing with the P800. It's all about improving things at this stage.'

Renderings of the P910 from the Sony Ericsson Design Centre, 2004

245

Text Messaging

It was just meant to be a tool for engineers to test the phone networks with. It was a shock, therefore, when customers started using SMS (short message service) technology to communicate with each other.

The phone companies had presumed that text messaging was too difficult and time-consuming to take off, and, in this perfect example of what really drives technology, there's something quite satisfying about how wrong they were.

Text messaging was the solution to a whole host of communication problems we never knew we had. Want to say something instantly, but don't want to have a conversation? Send a text! Just like email (another accidental revolution), it was a brilliant new way to keep in touch with people that fits easily and conveniently into our lives.

SMS continues to surprise, and as new applications for it emerge it is still rarely out of the headlines. During the tsunami disaster in Asia in 2004 text messages could get through where the signal was to weak to sustain a conversation, and proved invaluable in organizing aid distribution.

The virtual keyboard (VKB) designed by Priestman Goode, projects a keyboard onto any flat surface using laser for easier texting, 2004

Text Messaging

New languages evolved so we could cram as much meaning into as few lines of text as possible. Like voice calls, new codes of etiquette were established to determine what was appropriate to text and where. Annual texting championships have seen punching letters into a phone become quite a sport.

Text messaging has had its critics though. Parents worried that their children would lose the ability to write without weird emoticon brackets and numbers. Doctors in the United States worried about the health of the nation's thumbs. They were not designed for getting information into a system, they said, and people were surely at risk of developing painful conditions, while

health experts in Britain and Korea also reported an increase in complaints of sore thumbs among text-messaging enthusiasts. In Australia in 2003 there was even a national day of safe text during which people wore bandages on their thumbs while practising 'text-ercises'.

The risk wasn't so big as to become a real problem, although perhaps the world-champion fastest text messenger ought to bring out a thumb-fitness video.

The abbreviated language that appears in texts (nice 2 C U; 2geva 4eva; and so on) began in Japan, where text mixed with smileys and *kao-moji* – more complex face characters like (;_;) to express sadness or m(___)m,

Nokia 3220 with air-texting facility, 2004

The razor-toothed piranhas the genera Serrasalmus and the most ferocious freshwa In reality they seldom att

The most challenging sentence in the 2004 texting championships involved writing about piranhas without using any predictive facilities.

which translates as a little humble thing with its hands held up in meek apology – have been around for a while.

Japanese schoolgirls have since moved on to a secret code of characters called *gyaru-moji* – a mixture of Japanese syllables, mathematical symbols, numbers and Greek characters. Taking SMS to a more personal level, some prefer to jot down their message on a piece of paper, take a photo of it and send it as a picture message.

Another evolution of the idea was air-texting. The first handset to incorporate this new facility was Kurv, by Kyocera, which was launched in 2003 and targeted at the teenage market. Especially handy in nightclubs,

ocontrus are

fish in the world.

uman.

NOKIA

Connectivity

An apocryphal tale from the mid-1990s told of a man sitting on a train on the London Underground showing off and talking loudly into his phone. Phones were rare at the time and were often owned by yuppie characters such as this one. He's chatting on his huge black brick when an old lady further down the carriage suddenly clutches her chest and falls to the floor. She's having a heart attack. Another passenger, who has noticed our hero with a phone, asks to borrow it to call an ambulance. He has to admit to the carriage that he was lying, talking to no one, because you can't get a signal on the London Underground – yet. With very few exceptions, phone technology now enables people to contact others almost anywhere instantly. Once this process took days, even weeks, now it's as easy as one-two-three.

As cellular technology spreads to ever more remote corners, and WiFi wireless internet and VoIP add to our communication capabilities, the world has shrunk. And with Bluetooth and voice recognition we can do it all with no hands.

Perpetual contact has profoundly changed how people interact with each other – for better and for worse. People are now far less likely to commit themselves to fixed appointments. Options are kept

ng post by a roadside, Philips Telecommunications

as open as possible for a long as possible in the knowledge that arrangements can be confirmed or changed at any time until they actually happen. They are merely 'approximeetings'. We have wider and shallower networks of contacts than before: we have more friends and we socialize in larger groups more frequently, but we spend less meaningful time with any of them.

Time spent on transport, for example, once considered 'dead time', is now time spent making phone calls. Instead of having more time as a result of greater connectivity, we are greedy for more experiences, and as a result can find ourselves with less.

Such connectedness, however, has changed more than our social lives. It has saved thousands of people from poverty and illness. In developing countries, phones are helping the poor to conduct business more effectively, while hospitals all over the world have implemented text-message systems

Ericsson W800 Walkman phone, 2005

Connectivity

to remind people to take medication, communicate diagnoses or to convey advice.

For every improvement in connectivity, however, there's a new worry for security and privacy too. As long as your phone is switched on and has a signal you are not alone and you can be found. While this has its advantages if you're up a mountain and need rescuing, for example, it also has some serious pitfalls. The knowledge that help is just a phone call away can make people more reckless in their behaviour. Furthermore, there are huge security and privacy implications.

As more devices came with Bluetooth technology, a new pastime was invented: Bluejacking. Especially popular on trains where victims are captive, Bluejacking involves using a phone to search for any other phones that have Bluetooth switched

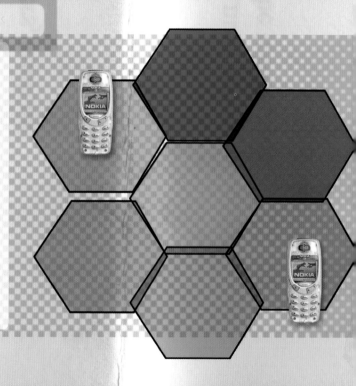

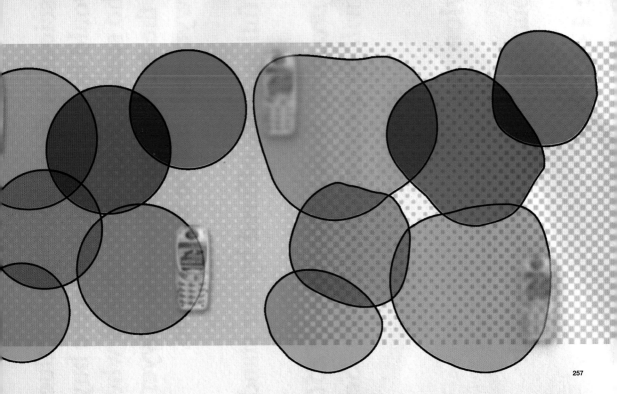

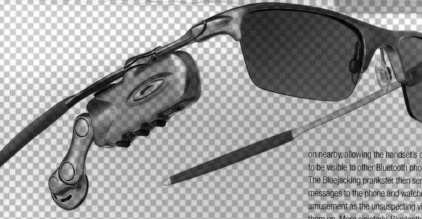

The Oakley Thump Bluetooth music player and hands-free kit on a pair of sunglasses, 2002

on nearby, allowing the handset's details to be visible to other Bluetooth phones. The Bluejacking prankster then sends witty messages to the phone and watches with amusement as the unsuspecting victim picks them up. More sinisterly, Bluetooth has also been used to send phone viruses.

Location-based tracking makes it harder to lie about your whereabouts, and phone networks are able to pinpoint the precise location of a handset to within ten metres or less. GPS chips and triangulation (bouncing signals off three phone masts to establish an exact set of coordinates) are popular with the military and police, with transport and logistics companies who want to know where their vehicles are and with companies who want to know where their employees are. Others see it as a very worrying development, as the scope for the

ology's misuse is enormous.
d it is not just shops, restaurants and
has beaming you special deals as you
the neighbourhood. There is more
tial for subscriber 'social' services,
as matchmaking or city guides and
Perpetual contact? It's hard being
, but one thing's certain, it's going to
arder.

Motorola Bluetooth-enabled motorcycle helmet, 2004

Dialtones
Technical realization diagram

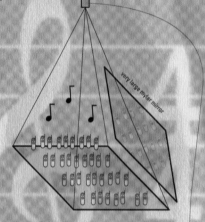

(D) Vertical video beamer, projects the performance computer's video display onto the audience itself

very large mylar mirror

microcell antenna, modified

participants' mobile phones

The audience is the orchestra. participants come pre-equipped with their own instruments, or borrow an instrument for the event.

The audience is the active score. when an instrument is activated, a spot of light is simultaneously cast onto it from the video projector, above. Instrumentalists may observe themselves in a 12-meter wide mirror, opposite. Participants also hold small keychain devices (E), which light up when near a ringing phone.

(A) Web-based registration terminals capture participant information and provide special seating tickets.

Database server manages records of phone numbers, seating locations, ringtones, and phone models.

(B) Performance computer (Windows2000 + OpenGL) with grid-based visual interface, permits conductor to dial audience phones. in deliberate patterns.

two 2Mbit E1 lines, connects phone calls. Up to 60 phone calls may be placed at the same instant.

Ringtones

Whether you want it to or not, your ringtone says a lot about you. More than clip-on covers, screensavers, and other accessories, ringtones have become the world's favourite way to express individuality through customizing a phone. From chart hits, TV theme tunes and animal noises on the bus to foghorns, car engines, carnal sound effects and smashing glass in the shopping centre, since the phenomenon began, the soundtrack to our world has changed dramatically.

While phones themselves might be a mark of civilization, many ringtones on the market undermine any such claim. It may be baffling to imagine what someone who wants their phone to sound like a drowning frog might be trying to communicate about phone calls.

themselves, but for companies sell them it doesn't matter. The phone ringtone business is worth billions of dollars.

Nokia are responsible for the ring-rage induced by the 'Gran Vals' series of beeps adapted from Fran Tarrega's composition that it issue on its phones in the 1990s. It was also the first company to enable SN – the technology that inspired Finn computer programmer Vesa-Matti Paananen to develop the software Harmonium, which he put on the internet for people to download fre charge. It allowed people to compo basic musical sequences on their phones and send them via SMS. W Paananen didn't profit, hundreds o others saw the potential.

Ringtones

However, the hideous series of single tinny notes was never really that exciting. But in 2002 polyphonic technology became widely available and phones began to sound like a tinny orchestra instead. Eventually polyphonic technology was eclipsed by the capacity to play CD-quality 'true tones', so phones could play just about any song or sound, if required.

Now record companies often market the ringtone as energetically as they do the CD single. Or it might be available with a code purchased with the disc, while artists take a generous percentage in royalties every time it's bought.

There is still scope for innovation, and novelties such as the 'breast-enlarging' ringtone from Hideto Tomabechi attracted over 10,000 downloads in the first week it was available. Tomabechi dubiously claimed to have developed a tune that would increase the breast measurements of those who listened to it using sounds that make the brain and body move subconsciously.

Tomabechi's ambitions for ringtones include ones that improve memory, increase attractiveness to the opposite sex, make hair sprout and help people to give up smoking. Playing on people's insecurities perhaps? No more than any other ringtone self-expression does, surely.

Camera Phones

After wireless voice, there was wireless text. And after wireless text, came wireless images. Camera phones created a third revolution in mobile communication.

Camera phones first appeared in Japan in 2000 when carrier J-Phone released what is claimed to be the first model, the J-SH04 by Sharp. J-Phone invested heavily in a camera-phone future and by 2002 it was apparent that this had paid off. The company's profits shot through the roof. The West – whose manufacture had been sceptically holding back plans for camera phones for a while – finally decided it wanted what J-Phone was having, and the revolution hit Europe and America about a year later.

Motorola T725, 2004

As with mobile phones, the value of mobile imaging was hard for many people to grasp at first. Why would anyone need to take miniature, poor-quality pictures when they could take proper ones with a proper camera?

Once they started to use them, however, camera phones quickly became ubiquitous and indispensable. People soon realized that, because they took their phones to far more places than they would take a camera, they had more occasions to take pictures. Furthermore, camera phones allowed people to share photos instantly. 'Life caching', recording your life and sharing it using digital media, became the new diary writing, and passers-by were at the frontline presenting local current affairs on web blogs and selling their images to the news. With a camera phone, anybody could be a reporter.

Taking and sharing pictures became an everyday thing for millions of people, and the quality of images got better and better. KDDI launched the first one-megapixel phone in 2003 (the A5401CA by Casio) and Samsung was first to offer five-megapixels – better than many 'proper' digital cameras in circulation – in 2004 with the S250 [p. 166].

While the new generation of phones was hugely popular, it could also be a major intrusion. As they became more prolific – with almost every new phone released incorporating a camera as standard by 2004 – concerns over privacy began to spread.

Camera Phones

...om opportunistic shots up skirts to ...erious corporate theft, camera phones ...ontained a lot of potential that could ...e abused.

Solutions were varied: Saudi Arabia ...anned them, although no one took ...uch notice. In South Korea, the ...overnment specified that camera ...hones should be designed to produce ...loud sound when a picture was ...ken. London-based watchdog Privacy ...ternational recommended a default ...ash to prevent people from taking ...overt pictures. One UK-based company, ...ensaura, launched the technology Safe ...aven, which disabled a phone's camera ...ithin a localized environment, such as a ...hanging room or a museum.

Many organizations, including schools, gyms, companies and government offices, banned them from their buildings for fear of sensitive information being snapped and leaked.

Protesters were left wondering why, in the rush to ban camera phones, few remembered that photography was the issue here. Stand-alone Cameras are not mentioned anywhere. It just proves that it's always the youngest one who gets the blame.

With the advent of 3G, these concerns grew, but the possibilities also became more interesting, as video phoning became a reality. You could film your friends, babies and the odd tourist attraction simply and easily and send clips via MMS. It also meant that the holy grail of future-gazers everywhere

Nokia PTG remote camera, the CCTV-style fixed camera sends a photo to your phone or email address by MMS each time it detects movement, 2005

Life blogging enables people to uploa
texts, images and video into a virtual

Camera Phones

s finally in sight: mobile video nferencing. Users could see and hear ch other in real or near real time. It was surprising, therefore, that video lling took a while to take off. Many ople found the idea of holding a device front of their faces hard to get used to. s often claimed that pornography is e handmaiden of new technology and, th so much potential for the industry to oloit, there were inevitably those who

worried that video phones would only contribute further to the fall of civilization as we knew it. As camera phones had already proved, however, the technology is often blamed unfairly, and once again phones simply brought more power to the people.

The Rise of 3G

For an outsider, the phone industry can seem like a mass of confusing acronyms. And never more so than when talk turns to third-generation technology, or 3G.

First there were wireless networks that used analogue technology to transmit voice; second-generation (2G) networks were digital, increasing sound quality and allowing data to be transmitted as well. The next step, awkwardly called 2.5G, allowed 'always-on connectivity'.

At this stage the Global System for Mobile Communications (GSM) was upgraded to the General Packet Radio Service (GPRS) making way for 3G capability. As well as allowing for a permanent data connection it also meant that users could be charged only for the services they accessed, not the duration of the connection. For 3G, Enhanced Data GSM Environment

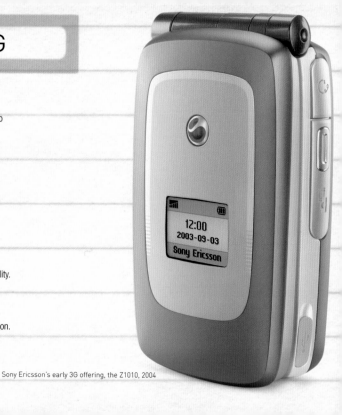

Sony Ericsson's early 3G offering, the Z1010, 2004

The Rise of 3G

(EDGE), enabled all this to happen much faster. EDGE, and competing standard Code Division Multiple Access (CDMA), promised extremely fast data transmission and much more capacity. Essentially 3G was like broadband for phones.

Before 3G, the networks were hugely over-burdened with their data loads, so carrier companies were extremely eager to provide this technology, and in 2003 European governments auctioned 3G licence fees for billions of euros. And they hadn't even built the networks yet.

3G was sold to consumers as the gateway to a world where phones could do practically anything from practically anywhere, super-fast and super-easily. Internet? Video? Music downloads? No problem.

Nokia 6630, 2004

Or there shouldn't have been, but it quickly became apparent that 3G technology was big on hype, short on delivery. It was portrayed as something vaguely necessary and life changing, but it was difficult to see how. And the extortionate costs involved as the companies tried to recoup the billions paid for licences weren't really justified by the ability to download video-clips 'on the go'.

It was not the first time that the phone industry had invented a technology before it had worked out what it would be used for. The race was on to find the 'killer application' for 3G but, as with phones themselves, carriers found it hard to force-feed people technology before they really wanted it.

We were told how wonderful it would be to experience the wonders of video calling. As if the idea hadn't been unappealing

Vodafone V603T, 2004

The Rise of 3G

enough when it was available over a landline and called video conferencing. As if the ability to make a video call while you are actually mobile — walking down the street (holding a phone in front of your face) — wasn't plain daft. It would only end in tears and dented lampposts, and we all knew it.

Other suggested applications included 'mobisodes' — miniature episodes of exciting made-for-mobile television dramas. But the idea of being able to watch them at the bus stop,

unsurprisingly, didn't send us scurrying for our wallets. We didn't buy any of it. Or many of the phones.

Another problem for 3G was that the trend in handsets was still that smaller equals better, but small phones don't support rich media delivery very effectively. When the first 3G phones emerged they didn't look especially desirable.

The mistake of 3G was that it was marketed as a new toy to maximize phone fun, for watching videos,

above and opposite NEC N901, 2004

Video messaging. Live it.

The Rise of 3G

downloading music and playing games, but it was also a building block for the future of the phone industry, and it could only be realized when the right services were available to make it useful. Although Japan's high-speed mobile internet service – NTT DoCoMo's i-Mode, which was introduced in early 1999 – was an immediate success, it was also the first experience many Japanese had of the internet, and the internet is indispensable, mobile or not. In Japan, people had been using their phones to receive news and stock prices, to email and to shop and bank online for years before the rest of the world got 3G. The real appreciation of 3G, however, goes far beyond this. With a fast, mobile connection to the internet it becomes possible not just to talk to anyone anywhere, but to see anywhere. Whether you see the video images taken by another phone or those by a stationary webcam or security camera, the possibilities are extreme. See the same view that your partner sees while you are apart, check the roads for traffic before you travel or scan the car park for a space before you get there. See if your date has arrived at the bar yet or, perhaps, if someone better has. They said the future was teleportation. We don't need it.

Directory

Designers/Manufacturers/Networks

3
www.three.co.uk

Ericsson
www.ericsson.com

Naoto Fukasawa
www.naotofukasawa.com

Handspring
www.handspring.com

Hulger
[formerly Pokia]
POKE
Biscuit Building
10 Redchurch Street
London, E2 7DD, UK
T +44 (0)20 7749 5359
F +44 (0)20 7749 5383
www.hulger.com

IDEO
www.ideo.com

Ilkone
www.ilkonetel.com

KDDI au
www.kddi.com/english

Kyocera
global.kyocera.com

LG Electronics
www.lge.com

Motorola
www.motorola.com

MTN Village Phone
www.mtnvillagephone.co.ug

NEC
www.nec.com

Nextel
www.nextel.com

Marc Newson
Paris Studio – Nimrod Sarl
19 rue Beranger
75003 Paris France
T +33 (0)1 44 78 87 30
F +33 (0)1 44 78 87 39
www.marc-newson.com

Nokia
www.nokia.com

NTT DoCoMo
www.nttdocomo.co.jp/english

O2
www.o2.com

Orange
www.orange.com

Pantech
www.pantech.com

PetsMobility Network, Inc.
10575 114th Street, Suite 2
Scottsdale
AZ 85259, USA
T +1 480 344 7724
www.petsmobility.com

RIM
www.blackberry.com

sung/Samsung Design Europe
samsung.com

u Sanada
applele.com

H!
coudal.com/shhhcards.pdf

m Telecom
phase-IV, Udyog vihar
aon-12201, India
1 124 510 8830/34
124 501 8117
shyamtelecom.com

ens
.siemens.com

SIMpill
PO Box 30451
Tokai, 7966
South Africa
www.on-cue.co.za

Sony
www.sony.net

Sony Ericsson
www.sonyericsson.com

Strapya
2-9-46 Oh-Big Building 3F Odawara
Kanagawa 250-0011, Japan
T +81 465 22 8064
F +81 465 22 8065
www.strapya.com

T Mobile
www.t-mobile.com

Toshiba
www.toshiba.com

Vertu
www.vertu.com

Vodafone
www.vodafone.com

Engadget
www.engadget.com

Jamster
www.jamster.com

Ringtonia
www.ringtonia.com

Slashphone
www.slashphone.net

Textually
www.textually.org

SMS Dictionary

abL	able	BaB	baby	CT	city	don8	donate
activ8	activate	BWD	backward	clubN	clubbing	dr$	dress
adr$	address	B&	band	congr@tul8	congratulate	dubL	double
al2gethr	altogether	Bcum	become	c%l	cool		
amazN	amazing	B4	before	koZ	cosy	evR	ever
anoyN	annoying	Bside	beside	cupL	couple	EZ	easy
aQr8	accurate	b%k	book	craZ	crazy	XTC	ecstasy
aw8	await	boC	bossy	cre8	create	FX	effects
aX	across	br&	brand	c%d	could	NRG	energy
ax$	access	b@RE	battery	communic8	communicate	MergNC	emergency
&	and	bl$	bless	compNE	company	MT	empty
R	are	brEZ	breezy			Nd	end
RTcl	article	bubLE	bubbly	D8	date	NtRtAn	entertain
LRG	allergy			Db8	debate	Scape	escape
@	at	KpabL	capable	dX	decks	Xample	example
@ak	attack	cLebr8	celebrate	DIAd	delayed	Xpect	expect
RTficL	artificial	cL	cell	DV8	deviate	XpNsiv	expensive
NE1	anyone	cNtrL	central	dinR	dinner	XperENs	experience
NE	any	centR	centre	duz	does		
		chocl@	chocolate	doN	doing		

280

fantaC	fantasy	gNRLE	generally	humR	humour	l&	land
favR	favour	gNR8	generate			l8	late
flavR	flavour	gNR8n	generation	LEgL	illegal	l8r	later
finL	final	gNtL	gentle	im@UR	immature	lytw8	lightweight
fl@	flat	glamRS	glamorous	im£ed	impounded	l%k	look
f%d	food	g%d	good	impr$	impress	l%P	loopy
f%T	footie	gradu8	graduate	NcreDbL	incredible	lotRE	lottery
4	for	gr8	great	infl8	inflate	luvr@	love rat
4N	foreign	gr%V	groovy	in2	into	lyN	lying
4plA	foreplay	grUsum	gruesome	ir8	irate		
4evR	forever	garNT	guarantee			mNE	many
4get	forget			jokN	joking	mRvLS	marvellous
4giv	forgive	hamRd	hammered	jLSE	jealousy	m8	mate
4nite	fortnight	h&D	handy	Gsus	Jesus	m@R	matter
4tun8ly	fortunately	hapN	happen	jRNE	journey	mEtN	meeting
4ward	forward	h8	hate	juC	juicy	mLO	mellow
frNZ	frenzy	hRt	heart	juvNil	juvenile	mLOD	melody
frNd	friend	heV	heavy			memRE	memory
		LO	hello	Xogram	kissogram	m$ge	message
		hiR	higher	QmQ@	kumquat	myt	might

SMS Dictionary

milNEM	millennium	pRticUlR	particular	remMbR	remember	T	tea
moV	movie	pRtnR	partner	resRv8n	reservation	t&	tanned
		pRT	party	respX	respects	teknoloG	technology
nSSRE	necessary	PpL	people	rStRNt	restaurant	tXt	text
nEd	need	pRsN	person	r%m	room	thnQ	thank you
neg@iv	negative	pRsNL	personal			thnX	thanks
nAbR	neighbour	pl&	planned	sAfT	safety	th@	that
nevR	never	pr@	pratt	s@RdA	saturday	thM	them
nyt	night	pr$	press	sX	sex	thN	then
nufN	nothing	puCk@	pussycat	shopN	shopping	2	to
no1	no one	prv8	private	sh%d	should	2dA	today
				sngL	single	2moro	tomorrow
ocup8N	occupation	qRL	quarrel	sk8	skate	2nite	tonight
ofN	often	qEZ	queasy	smL	smell	2wrd	toward
opN	open	qSTN	question	s%n	soon		
oper8N	operation	Q	queue	st&	stand	upd8	update
opR2n@E	opportunity			st8N	station	undRst&	understand
ovRdU	overdue			steD	steady	un4gtebL	unforgettable
ovR8d	overrated	r8	rate	str$	stress	unl$	unless
ovR	over	reD	ready	suspX	suspects		
		reCv	receive				

282

VDO	video	yS	yes
valU	value	U	you
vU	view	yr	your
vLNtine	valentine		
w8N	waiting	z%	zoo
w8	wait	zN	zen
wan2	want to		
wNt	went		
wN	when		
Y	why		
W/O	without		
1dRfL	wonderful		
w%d	would		

Technical Glossary

2G The second generation of cellular networks (analogue was the first generation). 2G is a generic term that includes **GSM**, **CDMA** and **TDMA** standards.

2.5G Enhanced versions of **2G** networks that also allow packet data services such as voice, for example **GPRS**.

3G Enhancement to a **2.5G** network offering much greater data speeds. For example **WCDMA** and **CDMA2000** standards.

airtime The time spent on a call.

analogue The old-fashioned method of modulating radio signals so that they can carry information.

battery A chargeable device used to power phones.

Bluetooth Named after Harald Bluetooth, a mid-tenth-century Danish King who united Denmark and Norway, Bluetooth allows wireless connection between devices such as mobile phones and laptops via short-wave radio signals.

call barring Setting a phone to prohibit specific incoming or outgoing phone calls.

call timer A feature on phones that keeps track of airtime minutes.

call divert A feature on phones that will divert incoming calls to another phone or voicemail.

call hold A feature that allows a caller to be made to wait, for example while another call is answered.

CSD Circuit Switched Data. Used to transmit data from a phone, although **GPRS** is generally faster and cheaper.

caller display Shows the incoming number and name on the phone.

CDMA Code Division Multiple Access. Digital cellular standard used in Japan, parts of the Far East and parts of the USA.

Technical Glossary

CDMA 2000 Also less catchily known as IMT-CDMA Multi-Carrier or 1xRTT, it is a **CDMA** version of a **3G** standard.

cellular A radio phone system in which a network of transmitters links the phone user to the public phone system. Each transmitter covers users in its own 'cell'.

CODEC Encodes and decodes (or compresses and decompresses) data for transmission, particularly sound and video files.

coverage The area in which a phone can make and receive calls, generally given as a percentage of the population that could use a particular network.

digital The new analogue: digital technology generates, stores and processes data in terms of little 0s and 1s. In mobile phones speech is converted into digital data, transmitted and then converted back to normal sound by the receiving phone.

dual band Describes mobile phones that can switch between different **GSM** frequency bands.

EDGE Enhanced Data GSM Environment. An air interface developed specifically to meet the bandwidth needs of **3G**. It is a faster version of **GSM** wireless service.

GPS Global Positioning System. A worldwide radio-navigation system formed from a constellation of satellites allowing users with suitable equipment to identify their geographic location.

GPRS General Packet Radio Service. Allows continuous connection to data networks at a high rate.

GSM Global System for Mobile Communication. A communication standard in three frequency bands: 900MHz, 1800MHz and 1900MHz.

hands-free Available as an accessory for most phones, a hands-free unit uses wires or **Bluetooth** to allow people to use phones without holding the handset to their ear. Good for drivers, and for looking like you're talking to yourself.

i-Mode NTT DoCoMo's mobile internet access system. Available in Japan long before Europe got **2.5G** or **3G**, i-Mode gave users the means to access the full internet at fast speeds on their phones.

Infrared A method of transferring data between devices that are positioned close to each other.

MMS Multimedia Message Service. Two-way messaging of text, images, sound and video.

Phone Book The list of names and telephone numbers stored in a phone's memory or on its **SIM card**. Also a brilliant publication written by Henrietta Thompson, published by Thames & Hudson, that everyone who has a phone would do well to buy.

predictive text input Allows users enter text on a phone faster by predic what words are being written. Invariabl baffling for first time users.

pre-pay/pay-as-you-go Term use for no-contract, no-rental-charge serv where you buy credit for calls in advar Each network has its own pre-pay ser

Push-to-Talk (PTT) A two-way communication service that works like a walkie-talkie. A normal phone call is full-duplex, meaning both parties can hear each other at the same time. PTT half-duplex, which means communica can only travel in one direction at any given moment. PTT requires the perso speaking to press a button while talkir and then release it when they are finis The listener then presses his button to respond. This way the system knows which direction the signal should be travelling in. Over and out.

Radio Frequency Identification. A ...ology that uses the electromagnetic ...ctrostatic coupling in the radio ...ency portion of the electromagnetic ...rum to uniquely identify an object, ...al or person. RFID tags are the new ...des – with the advantage of not ...ring direct contact or line-of-sight ...ning.

...ning The ability for a customer to ...ss services and make calls even ...travelling outside the geographical ...age area of their home network by ...another network available in that ...in instead.

...card Subscriber Identity Module. A ...card fitted inside a phone that stores ...ser's identity (PIN), information added ...e user and text messages.

...rtphone The term is used describe ...one with special computer-enabled features not previously associated with telephones, for example email, online banking and remote data transfer.

SMS Short Message Service. Two-way text messaging service.

SMTP Simple Mail Transfer Protocol. A protocol for sending email messages between servers and from client to server.

splashpad A vessel that charges a mobile device that is placed within it, without wires.

standby time The amount of time a battery at full charge will keep a phone running when not making or receiving calls.

talk time The amount of time a battery at full charge will keep a phone running if being used continuously.

TDMA A cellular standard that divides each cellular channel into three time slots, increasing the amount of data that can be carried.

tri band Phones that are able to operate on three **GSM** frequency bands: GSM900, GSM1800 and GSM1900.

UMTS Universal Mobile Telephone System. Another name for 3G.

voicemail If a phone user is unavailable or the phone is switched off, callers can leave a message on voicemail to be retrieved later.

VoIP Speech is packaged in data packets and then transferred over the internet, meaning that voice calls can be made much more cheaply, or even free.

voice recognition Enables phone functions to be controlled with voice commands.

WAP Wireless Application Protocol. A way for mobile phones to access services and information.

WCDMA Wideband **CDMA** – a **3G** standard for more data through-put.

Wi-Fi Stands for 'wireless fidelity', and is the term for secure wireless internet technology. The term was created by an organization called the Wi-Fi Alliance, which oversees tests that certify product inter-operability.

Acknowledgments

There are many people without whose help this book would not have been possible. Charlie Sorrel, gadget guru, ace designer and tireless conjurer of images, heads up the list, along with the extremely supportive and understanding editors and staff at Thames & Hudson, in particular Andrew Sanigar and Alice Park.

Thanks also to my editor at *Blueprint*, Vicky Richardson, for her support, and to Junko Fuwa for her amazing contacts, Japanese and sheer helpfulness at every turn. Clive Grinyer at Orange, Kevin McCullagh at Plan, Crispin Jones, IDEO's Graham Pullin, Michael Carrol and Seymourpowell's Richard Seymour were also invaluable for their advice at various stages. Thanks also to Alastair Curtis, Frank Nuovo, Bill Sermon and Claire Curtis at Nokia, to Gavin Spicer at Chocolate, Mark Kean at Karla Otto, and Andrew Ferguson at Purple. Also to all at Motorola, Firefly and the Fish Can Sing, to Michael Henriksson at Sony Ericsson, to Clive Goodwin at Samsung, Sheryl Sietz at Bite, Patsy Youngstein and Marc at Marc Newson and Elaine Quinn at three. In Japan thanks to Satoshi Sunahara at KDDI and Tomonari Higuchi at Strapya.

I would also like to thank the various manufacturers, service providers and other sources for the illustrations listed below. Every effort has been made to trace the copyright holders of the images contained in this book, and we apologise in advance for any unintentional omissions. We would be pleased to insert the appropriate acknowledgment in any subsequent edition of this publication.

BEDD: 202 (right), 13 (left + right). © BEDD 2005; Channel 4 Television/ Absolutely Productions: 264–65. © Channel Four Television Corporation MCMXCIX; Diva Berlin: 86. © 2002–2005 Diva Berlin; DoCoMo: 56, 58, 59. © 2005 NTT DoCoMo Inc; Clive Grinyer: 115; Hulger: 2, 4 (top), 81. © 2005 Hulger; IDEO: 134–35, 142–43, 200. © 2002 IDEO; Ilkone: 52, 53. © 2005 Ilkone Mobile Telecommunications; KDDI: 4–5 (bottom), 108 (bottom left), 150–51, 168–69, 222–27. © 2003–2005 KDDI Corporation; KYOCERA: 167. © KYOCERA Corporation; LG Electronics: 51, 92 (right), 170, 171, 196 (left). © 2001–2005 LG Electronics. Illustrations for pages 92 and 171 supplied courtesy of HealthPia America; Phil Marso: 37 (left + right). © 2004–2005 Phil Marso; Motorola: 11, 15, 20 (right), 24–25, 80, 87, 90–91, 96 (far left + 3rd from left), 97 (2nd from left), 98, 100, 108–109, 109 (right), 110–11, 114, 116–17, 122–23, 146–47, 155, 156, 172–73, 201 (left), 205 (left), 236–41, 259, 264. © Motorola Inc; MTN Village Project: 18, 45, 47, 48, 64, 65. © 2005 Grameen Foundation USA; murmur: 28, 29. © 2005 [murmur]; MyMo: 14. © 2005 MyMo; NEC: 66, 67, 68, 69, 97 (far right), 157, 165, 274 (left), 275 (right). © 1997–2005 NEC Electronics; Nokia: 8–9, 26, 27, 40, 41, 84, 95 (left), 96 (2nd from left + right), 97 (left), 101, 109 (bottom left), 118–19, 120–21, 124–25, 126–27, 128, 130–31, 132–33, 164, 174–75, 232–35, 248, 251, 256–57, 266–67, 269, 272. © Nokia 2005; Jamie O'Brien: 112–13; Oakley: 258. © 1998–2005 Oakley Inc; OneWorld: 44, 49. © 2005 OneWorld. Photographs courtesy of Peter Armstrong, OneWorld; Orange: 54, 55, 154. © 2005 Orange Personal Communications Ltd; PalmOne: 138–39, 142–43. © palmOne Inc; Pantech: 62, 180–81. © 2002–2005 Pantech Co., Ltd; RIM BlackBerry: 23 (right), 108 (top left), 136–37, 144–45. © 2005 Research in Motion Limited; Samsung: 19, 63, 88, 106, 107, 148, 166, 182–83, 216–21. © Samsung Electronics Co LTD Seoul Korea. Photographs on pages 106–7 by John Reynol[...] Isamu Sanada: 176–77. © 2005 Isamu Sanada; SHHH!: 22, 23 (left). © 2005 Coudal Partners In[...] Siemens: 85, 88 (right), 97 (3rd f[...] left), 152–53. © Siemens AG 20[...] SIMPill: 46. © 2005 SIMPill; SK Telecom: 61. © 2005 SK Telecon[...] Sony Ericsson: 16, 42, 129, 149, 160–61, 162–63, 194 (right), 19[...] 242–45, 255, 268, 270–71. © S[...] Ericsson Mobile Communications[...] Charlie Sorrel: 1, 71, 73, 54–75, 78, 79, 205; T-Mobile USA: 158. 2002–2005 T-Mobile USA, Inc; V[...] 140–141, 228–31. © Vertu 200[...] Vitaphone: 92–93. © 2005 Vitapl[...] VKB: 246–47. © 2002–2005 VK[...] Vodafone: 10, 32, 57, 82, 83, 89, (left), 159, 178–79, 208, 209, 21[...] 211, 273, 274, 275 (centre), 276[...] 2001–2005 Vodafone Group.

3G **267, 270–77**

accelerometer **182**
accessories **260**
address books **202**
Africa **44–49**
Ahlgren, Erik **162–65**
air-texting **248, 250**
Amena **35**
antennae **99, 214**
Apple **177**
Appling, George **145**
Arabic **19**
Arnholt, Chris **236**
Ashall, Peter **229**
Asia **56–71**
Australasia **70–75**
Australia **17, 70–75**

batteries **99, 214, 208**
Beckham, David **30, 83, 89**
Belgium **21**
Berlin Diva **86**
Berlusconi, Silvio **34**
Blueprint magazine **168**
Bluetooth **11, 80, 90, 214, 252, 256, 258, 259**
Burton **11, 80, 90–91**

business cards **199**

Cai Ling **68**
camera phones **50, 264–69**
cameras **196**
Canada **26**
candybar **99**
cash **203**
Casio:
  A5401CA **265**
  Casio G-shock **223**
CCTV **184, 267**
CDMA **104, 272**
cell phone **17, 22**
China **19, 66–69**
clocks **200**
CODEC **214**
context-aware phones **189–90**
Cooper, Martin **99, 116**
Croatia **17**
Curtis, Alastair **233**

Dancall **115**
Danger Research **158**
Darkstar **227**
diabetic phone **93, 171**
diaries **202**
drunken phone calls **70**

EDGE **272**
email **136**
environment impact **207–11**
Esperanto **19**
Europe **30–43**

fingerprint recognition **62, 180**
Finkbeiner, Tanja **232**
Finland **19, 40**
France **36**
Freeplay **208**
fuel cell **100**
Fukasawa, Naoto **150, 222, 223–25**

games consoles **164, 204**
Germany **19, 38–39**
GPRS **270**
GPS **258**
Grameen Foundation **46**
Gran Vals **260**
GSM **21, 270**

Handspring Treo **138–39**
Handy **19**
Harmonium **260**
Healthpia **171**
Helsinki **40**
Henriksson, Michael **244**

Higashiizumi, Ichiro **226**
HiPod **176–77**
Holland **17**
Hulger **81**
Hungary **17**

IDEO:
  Kiss Communicator **134–35**
  Social Mobiles **142–43**
Ilkone **50, 52–53**
i-Mode **56, 277**
India **64**
Infrared **214**
internet **104**
iPod **177**
Israel **21, 54**
Italy **32**
Iwasaki, Ichiro **224, 226**

J-Phone **264**
Japan **19, 56–59, 277**
Jellicoe, Roger **236**
Jensfelt, Henrik **242**
Joly, Dom **32**

KDDI **6, 57, 265**;
  Apollo /Apollo 02 **226**
  au Project **222–27**
  Grappa /Grappa 002 **226**

Higashiizumi, Ichiro **226**
Infobar **150–51, 222–24**
Ishicoro **224**
talby **168–69, 227**
W11K **226**
keitai **56**
Kenya **48–49**
keypad **214**
keys **197**
Krollop, Rudi **116–17**
Kyocera **93**;
  Kurv **250**
  TU-KA S Phone **167**

landline phones **201**
LCD **208, 214**
Lee, Kun-Hee **216**
LG Electronics:
  8110 **170**
  F7100 Qiblah phone **50, 51**
  KP8400 **93, 171**
life blogging **268**

Madrid **35**
The Matrix **133, 166**
Mecca **8**
Melnikov, Ivan **43**
microphone **214**
Middle East, The **50–55**
MMS **267**

Mobira Talkman **100, 118, 118–19**
mobisodes **274**
Moo-Hyun, Roh **62**
Motorola **6, 25, 91, 100, 259**;
  3G **156**
  A1000 **156**
  A380 **156**
  Baby Phat phone **87**
  collaboration with Burton **11, 80, 90–91**
  DynaTAC **98, 99, 116–17**
  SCR-100 **110–11**
  StarTAC **122–23**
  T725 **264**
  Transportable **114**
  V3 RAZR **172–73, 236–241**
  V70 **146–47**
MP3 **205**
MTN villagePhone **46**
Murmur **28–35**
Muslims **8, 50–53**
MyMo **94**
NEC:
  3G **157**
  e606 **157**
  N620 **67**
  N900 **68–69**

Index

N900i **157**
N901 **274**
N910 **66, 67**
N940 **165**
Newson, Marc **168–69, 227**
New Zealand, **21**
Nextel Push to Talk **155**
Nikaido, Takahashi **223**
Nokia **6, 19, 40, 100–1, 140, 229, 260**;
  2110 **120–21**
  3210 **130**
  3220 **248**
  3310 **131**
  6110 **124**
  6630 **272**
  7110 **133**
  7200 **232**
  7260 **232, 234, 235**
  7270 **232, 234**
  7280 **174–75, 232**
  8110 **125**
  8210 **132, 232**
  8810 **126**
  8850 **127**
  9000 Communicator **128**
  Cityman **118–19**
  Hello Kitty **95**
  N-Gage **164**
  Nokia Fashion **232–35**

PTG remote camera **267**
North America **22–29**
North Korea **60**
Norway **19**
NTT DoCoMo **6, 56, 277**;
  Eggy **59**
  i-Mode **56**
  Pocket Post Pet **58**
  Wristomo **56**
Nuovo, Frank **140, 229–31**

Oakley Thump **258**
Orange **54**;
  SPV **154**
  SPV C500 **154**
  SPV E100 **154**

Paananen, Vesa-Matti **260**
Pantech GI100 **62, 180**
pelephone **55**
PetsCell **22, 77**
Philippines **19**
Philips Telecommunications **253**
phone throwing **38**
Poland **19**
polyphonic **263**
Portable, le **36**
Portugal **17**
Pringle **84**

Push-to-talk (PTT) **24, 25, 155**
Pullin, Graham **142**

Qur'an **52**

recycling **208, 211**
remote controls **198**
RFID **58**
RF Unit **214**
RIM BlackBerry **136–37**
ringtones **32, 260**
Russia **21, 42**

Samsung **106**;
  and Diane von Furstenberg **88**
  Hera **216–21**
  N270 (*Matrix* phone) **107, 166**
  S250 **63, 166**
  SCH-S310 **182–83**
  SH-700 **216**
  T100 **148**
Samsung Design Europe **106, 217**
Sanadu, Isamu **176–77**
Saudi Arabia **50, 267**
Science and Mechanics **102**
screensavers **260**
Senegal **44**

Sensaura **267**
Seymour Powell Forecasting **217**
Sharp:
  J-SH04 **264**
  Vodafone live! **159**
Shoe phone **98, 117**
Shyam Telecom **64**
Siemens:
  and Escada **85, 88**
  SL55/65 **152–53**
  Xelibri **86, 144–45**
SIM **214**
SIMpill **46, 48**
Sky IM-7400 **61**
smoking **36, 94**
SMS *see* text messaging
Society for Hand Held Hushing (SHHH!) **23**
Sony CMD Z1 **129**
Sony Ericsson **6**;
  P-Series **242–47**
  P800 **42, 160–61, 242**
  P900 **242, 243**
  P910 **242, 245**
  P910i **242**
  T300 **149**
  T610 **162–63**
  W800 (Walkman phone) **255**
  Z1010 **270**

South Korea **60, 267**
South Africa **46**
Spain **19, 34**
squillo **34**
Sri Lanka **19**
Star Trek Communicator **10, 112–13, 122**
Sweden **17, 19, 42**
Symbian **160**

T-Mobile Sidekick **158**
Tarrega, Francisco **260**
Telecom Italia Mobile **33**
teleportation **277**
Tellumat Communications **48**
terrorism **10**
text messaging **12, 36, 38, 104, 246–51, 260**
Tomabechi, Hideto **263**
Toshiba V602T **178–79**
Trigger Happy TV **32**
true tone **263**
tsunami **246**
Turkey **19**

Uganda **46**
UK **17, 30–32**
USA **22–25**

Vatican, The **33**
vending machines **68**
Vertu **228–231**;
  Ascent **230**
  Signature **140, 229**
vibration **123**
video calling **267–69, 27**
Virgin **70**
Vocera **113**
Vodafone:
  Vodafone live! **35, 82–89, 159, 275**
  V603T **273**
  Toshiba V602T **178**
VoIP **201, 252**

WAP **104, 133**
watches **200**
wearable **193**
WiFi **252**
Williamson, Matthew **84**
Wise, Ernie **30, 32**
Wristomo **56**

Xpress-on covers **84, 99, 131, 132**

Zingo Taxis **30**